Tributes
Volume 15

Hues of Philosophy
Essays in Memory of Ruth Manor

Volume 7
Dialogues, Logics and Other Strange Things.
Essays in Honour of Shahid Rahman.
Cédric Dégremont, Laurent Keiff and Helge Rückert, eds.

Volume 8
Logos and Language.
Essays in Honour of Julius Moravcsik.
Dagfinn Follesdal and John Woods, eds.

Volume 9
Acts of Knowledge: History, Philosophy and Logic.
Essays dedicated to Göran Sundholm
Giuseppe Primiero and Shahid Rahman, eds.

Volume 10
Witnessed Years. Essays in Honor of Petr Hájek
Petr Cintula, Zuzana Haniková and Vítězslav Švejdar, eds.

Volume 11
Heuristics, Probability and Causality. A Tribute to Judea Pearl
Rina Dechter, Hector Geffner and Joseph Y. Halpern, eds.

Volume 12
Dialectics, Dialogue and Argumentation. An Examination of Douglas Walton's Theories of Reasoning and Argument
Chris Reed and Christoher W. Tindale, eds.

Volume 13
Proofs, Categories and Computations. Essays in Honour of Grigori Mints
Solomon Feferman and Wilfried Sieg, eds.

Volume 14
Construction. Festschrift for Gerhard Heinzmann
Solomon Feferman and Wilfried Sieg, eds.

Volume 15
Hues of Philosophy. Essays in Memory of Ruth Manor
Anat Biletzki, ed.

Tributes Series Editor
Dov Gabbay   dov.gabbay@kcl.ac.uk

Hues of Philosophy
Essays in Memory of Ruth Manor

edited by
Anat Biletzki

© Individual author and College Publications 2010. All rights reserved.

ISBN 978-1-904987-76-5

College Publications
Scientific Director: Dov Gabbay
Managing Director: Jane Spurr
Department of Computer Science
King's College London, Strand, London WC2R 2LS, UK

http://www.collegepublications.co.uk

Original cover design by orchid creative   www.orchidcreative.co.uk
Printed by Lightning Source, Milton Keynes, UK

All rights reserved. No part of this publication may be reproduced, stored in a retrieval system or transmitted in any form, or by any means, electronic, mechanical, photocopying, recording or otherwise without prior permission, in writing, from the publisher.

Contents

List of contributors	vii
Preface	ix
1. **What the Bald Man Can Tell Us** JAAKKO HINTIKKA	1
2. **Tonk - A Full Mathematical Solution** ARNON AVRON	17
3. **Davidson's Notion of Supervenience** ORON SHAGRIR	43
4. **Scientific Theory and Natural Language, Holism, and Measurement** ELI DRESNER	59
5. **Ornamentality in the New Media** ERAN GUTER	83
6. **The Imaginary Pieces** AMNON WOLMAN	97
7. **Lonely Beating: Wittgenstein's Automaton and the Drums of War** ANAT MATAR	117
8. **Living with Paradoxes** ANAT BILETZKI	131

Contributors

Arnon Avron is Professor of Computer Science at Tel Aviv University. His main interests are in mathematical logic, logic in computer science, philosophical logic (especially non-classical logics), proof theory, automated reasoning, foundations of mathematics, and philosophy of mathematics and logic. He has published articles in several journals including *Journal of Symbolic Logic*, *Journal of Philosophical Logic*, *Notre Dame Journal of Formal Logic*, *Information and Computation*, and has contributed chapters to, among others, *Handbook of Philosophical Logic* (Kluwer, 2002) and *Towards Mathematical Philosophy* (Springer, 2009).

Anat Biletzki teaches philosophy at Tel Aviv University and is Albert Schweitzer Professor of Philosophy at Quinnipiac University. Her interests travel from logic and philosophy of language to political philosophy and human rights. She is the author of *Paradoxes* (Defense Ministry Publications, 1996), *Talking Wolves* (Kluwer, 1997), *What Is Logic* (Defense Ministry Publications, 2002), and *(Over)Interpreting Wittgenstein* (Kluwer, 2003).

Eli Dresner received his PhD in Logic and Methodology of Science from the University of California at Berkeley (1998), and is currently a senior lecturer in philosophy and communication at Tel Aviv University. His research interests are in the philosophy of language, logic and the philosophy of computing, where, in his recent work, he has been applying measurement-theoretic conceptualization to questions in formal semantics and the theory of meaning. Some of his work has been published in *Synthese*, *Linguistics and Philosophy*, *The Journal of Philosophical Logic* and *Minds and Machines*.

Eran Guter is Senior Lecturer in Philosophy at the Max Stern Academic College of Jezreel Valley, and Researcher at the Department of Philosophy, University of Haifa, Israel. He is also affiliated with the Sammy Ofer School of Communications, Interdisciplinary Center Herzliya, Israel. His current research focuses on new media aesthetics and on Wittgenstein's philosophy of music. He is the author of *Aesthetics A-Z* (Edinburgh UP, 2010). He has published articles in the *British Journal of Aesthetics*, and in specialized volumes such as *Television: Aesthetic Reflections* (Peter Lang, 2002) and *Interactive Wittgenstein* (Springer, 2010).

Jaakko Hintikka is professor of philosophy at Boston University. Author of over 30 books, he is the main architect of game-theoretical semantics and of the interrogative approach to inquiry, and also one of the architects of distributive normal forms, possible-worlds semantics, tree methods, infinitely deep logics, and the present-day theory of inductive generalization. Five volumes of his *Selected Papers* (Kluwer Academic) appeared in 1996-2003 and in 2006 as the volume *The Philosophy of Jaakko Hintikka* was published in the series *Library of Living Philosophers*.

Anat Matar is senior lecturer of philosophy at Tel Aviv University. She is the author of *From Dummett's Philosophical Perspective* (de Gruyter, 1997) and *Modernism and the Language of Philosophy* (Routledge, 2006). Matar has been an anti-occupation activist for many years, in particular in the refusal movement. She is co-founder and Chair of the Israeli Committee for the Palestinian Prisoners and serves in the steering-committee of Who Profits? – Exposing the Israeli Occupation Industry.

Oron Shagrir is Associate Professor of Philosophy and Cognitive Science at the Hebrew University of Jerusalem. He works on various topics in philosophy of mind, foundations of cognitive science, and history and philosophy of computing. He has published articles in *Mind*, *Philosophical Studies*, *Philosophy and Phenomenological Research*, *Synthese*, *Philosophy of Science*, and others.

Amnon Wolman is a composer and sound artist who creates music for diversified forces of musicians grounded in an essential interest in experimentation and a belief that music as an art form expresses many dissimilar ideas of beauty. He previously taught at Northwestern University, Brooklyn College and the Graduate Center at CUNY. He is currently the director of the School of Music Education at Levinsky College in Tel Aviv, the artistic director of *Ensemble Musica Nova*, Tel Aviv, and teaches at Tel Aviv University.

Preface

The articles in this volume are the fruit of a conference convened in November 2006, in memory of Ruth Manor on the first *yahrzeit* of her untimely death in November 2005.

Ruth Manor, whom everyone knew as Ruru, was, early and professionally, a logician. Trained in Jerusalem and Pittsburgh, she studied with the "greats" of the 1960s in these respective locales—Yehoshua Bar-Hillel, Nuel Belnap, and Nicholas Rescher. Her earliest publications, co-written with Rescher, accordingly sported formal-sounding titles: "Modal Elaborations of Propositional Logics" or "On Inferences from Inconsistent Premises", and this gave rise to the Rescher-Manor Mechanism still used in logic today. But it is, indeed, while perusing her articles and books, that one immediately intuits the amazing trajectory that her work took in the ensuing four decades – from that preliminary proclivity for formal logic, to the (still formalistically-oriented) philosophy of language, to a more informal focus—in logic on fallacies and in the philosophy of language on pragmatics, then to rhetoric, and finally to ethics. Her articles evidence this route when she writes, early in her career, on "A Semantic Analysis of Conditional Assertion", then "An Analysis of a Speech", moving to the seminal "A Language for Questions and Answers" and the resulting problematic discussions on both the logic and the language of questions and answers. It comes almost as a shock that the last item in this list of articles is "Ethicians' Ethics"; in the meanwhile, between formal, mathematical logic and ethics she stops to think and write "On the Overlap between Semantics and Pragmatics", on "Solving the Heap", and uncannily, though still consistently, "Towards a Solution of the Middle East Conflict". The books are no different: The first book she collaborated on was, of course, true to logical character—*Studies in Modality*; the last, also typically collaborative (with her father, a renowned physiologist) was, of course, faithful to her growing ethical interests and angsts—*Doctors' Errors and Mistakes of Medicine – Must Healthcare Deteriorate?*

Such wide-ranging yet still profound versatility is difficult to impart—yet that is precisely what was attempted at the memorial conference, and is here reproduced. On the one hand, every article in this volume might merit a distinct disciplinary label; on the other hand, there is a clear—clear to those who knew and worked with her—Manorian silk thread connecting these articles. Jaakko Hintikka, with "What the Bald Man Can Tell Us", starts off the discussion with her abiding interest, from beginning to end—paradoxes—concentrating on the heap paradox, which she "solved" in her last publication. Arnon Avron, who worked with Manor on mathematical logic, appropriately provides, in Chapter 2, a mathematically formalistic article: "Tonk – a Full Mathematical Solution". In Chapter 3 Oron Shagrir continues the formalistic bent, but now with the more linguistic, dare I say more philosophical, concept which he adumbrates in "Davidson's Notion of Supervenience". Eli Dresner, in "Scientific Theory and Natural Language, Holism, and Measurement", addresses meaning holism and the relation between natural language and science; although, in contrast to the first three articles, this one does not house any formulas, it is no less meticulous for that, explaining the interaction between natural language and scientific theory via the rigorous idea of measurement. In the next chapter Manor's move to language, communication, and even rhetoric is reflected beautifully in Eran Guter's article "Ornamentality in the New Media". Her informal yet thoroughly intellectual perspective comes to bear in Chapter 6, Amnon Wolman's "The Imaginary Pieces"—a truly personal tribute to her always inspired and inspiring engagement with art and music. And her aggrieved, yet creative dealings with the ethical and political challenges of present day Israel are echoed in Chapter 7, in Anat Matar's "Lonely Beating: Wittgenstein's Automaton and the Drums of War". Chapter 8 closes a human chronological, scholarly circle, returning to paradox in Anat Biletzki's "Living with Paradoxes".

The links between the questions and issues that occupy the articles in the book came naturally to Manor. She was gifted in analyzing, that is to say, categorizing and compartmentalizing the different levels and aspects of our thought. Yet she exemplified the kind of understanding that Wittgenstein labeled as "seeing connexions". This was in no way a merely theoretical, abstract, or disembodied linkage between concepts alone. It was, rather, a form of life in which Manor connected philosophy with praxis, recognizing the paradoxes that inhere in both and insisting on fairness—in thought and in action—as the ultimate yardstick. We learned from Ruru how to live diverse though consistent, exacting though open-to-anything, logical though paradoxical, critical though charitable, troubled though ethical, philosophically exciting lives.

Anat Biletzki
Tel Aviv, 2010

1

What the Bald Man Can Tell Us

JAAKKO HINTIKKA

Dedicated to the memory of Ruth Manor

By speaking of the bald man, I am of course referring to the most clear cut of the paradoxes of vagueness, the sorites paradox. Or, strictly speaking, I am referring to one of the dramatizations of this paradox. This case is nevertheless fully representative of the general issues involved. (For the sorites paradox in general, see e.g. Keefe and Smith 1987 or Sainsbury 1995, ch.2.) The allegedly paradoxical argument is well known. It might be formulated as follows:

(P.1) Premise one: A man with no hairs is bald.

(P.2) Premise two: If a man with n hairs is bald, then so is a man with n + 1 hairs. (Colloquially speaking, adding a single hair will not rescue him from baldness!)

(C) Conclusion: No matter how many hairs a man has, he is bald.

The paradox lies in the fact that the premises (P.1) and (P.2) seem to be unproblematically true but the conclusion (C) false. Hence it is reasonable to have a closer look at the logical form of the argument that gives rise to the conclusion.

Hues of Philosophy.
Anat Biletzki (ed.).
Copyright © 2010.

Now the logical form of the statements involved here seems to be the following (using the conventional first-order notation):

(P.1)* $(\forall x)(H(x,0) \supset B(x))$
(P.2)* $(\forall n)(\forall x)(H(x,n) \supset B(x)) \supset (\forall x)(H(x, n+1) \supset B(x))$
(C)* $(\forall n)(\forall x)(H(x,n) \supset B(x))$

These in turn exemplify the members of the schema of *mathematical induction*.

(P.1)** $A(0)$
(P.2)** $(\forall n)(A(n)) \supset A(n+1))$
(C)** $(\forall n)A(n)$

The inferences from (P.1)* – (P.2)* to (C)* and from (P.1)** – (P.2)** to (C)** are valid in the usual first-order logic plus the principle of mathematical induction. Since neither of these inferences seems to be subject to doubt, we seem to have a serious paradox facing us. As was pointed out, the premises (P.1) and (P.2) seem true while the conclusion (C) is a (bald?) falsehood. Yet the inference turns out to be valid when it is construed in an apparently completely unproblematic way. As was pointed out, the inferences to (C)* and to (C)** from their respective premises are instances of mathematical induction. This mode of reasoning is thus involved in the problems of vagueness, and has in fact been discussed in connection with these problems. But it seems to be beyond any reasonable doubt, and in fact no serious doubts seem to have been leveled at it.

If the reasoning involved in the paradox seems so unproblematic, how can the paradox be explained away? What can the bald man tell us? Here comes the first and quite possibly foremost strategic moral that can be elicited from discussions of the different paradoxes such as the sorites paradox and the paradox of the liar. An enormous amount of inspiration and perspiration has been devoted to sundry "solutions" of the paradoxes. It is nevertheless by this time eminently clear that the only way of clearing up such major paradoxes is a deeper analysis of the basic concepts figuring in them. To discuss the paradoxes as separate puzzles without digging deeper into their sources in logic and semantics is not much more instructive than to solve crossword puzzles. The difficulty of the paradoxes is reflected in the fact that this proposed deeper analyses forces us to take a long hard look at the questions that underlie all logic. In the case of the sorites paradox the basic concepts that hold the key to understanding the paradox are *negation* and *mathematical induction*. (For negation, see Hintikka 2002.)

One way of coming to see this is to look at the attempts that have been made to "solve" the paradox. What are they like? One's first intuitive reaction to the puzzle of the bald man is undoubtedly to say: But baldness is not a black-or-white concept. In some cases—for some numbers of hairs—it does not make sense to say that a man is bald or that he is not. Technically, this is expressed by saying that the predicate "bald" exhibits *truth value gaps.* Surely that is why we do not really have to worry about the paradox.

This line of thought has given rise to attempts to deal with the paradox by postulating truth value gaps for predicates like "bald". In other words, it is assumed that our basic nonlogical predicates are not defined for some argument values, e.g. that $P(x)$ is not defined for the value b. Then $P(b)$ is neither true nor false. In the case of the paradox dealt with here, the vague concept will obviously be the notion of baldness.

Unfortunately but perhaps not unexpectedly, this line of thought does not yield a genuine solution to the paradox, at least not without a great deal of further ado. This further work has among other things led to the theory of supervaluations. Even without discussing the details of such theories, it is safe to say that they have not led to a definitive solution of the sorites paradox in spite of the plausibility of the idea of truth value gaps. Without going into any details, a sample difficulty will hopefully give you a sense of the difficulties here. Suppose that we allow the predicate "bald" to have truth value gaps. But let us then consider the predicate $B^*(n)$ that says that a man with n hairs is either bald or neither-bald-nor-not-bald. Then an obvious variant of the inductive argument above apparently proves that no matter how many hairs a man has, he is either bald or neither-bald-nor-not-bald, which is as absurd a conclusion as the original one.

There does not seem to be any way out of this predicament. The postulation of truth value gaps was intended to eliminate any sharp jump from bald to not bald by addition of one hair. But the very same postulation creates a similar sharp jump from cases of baldness to cases of neither-bald-nor-not-bald, thus re-creating the paradox.

Thus we have here a second paradox in our hands. It lies in the fact that the eminently plausible truth value gaps idea does not immediately solve the sorites paradox. Now the reason for this second paradox was already diagnosed above. It lies in reality in the superficial attitude to what the logic is that is involved in the sorites paradox. We must dig into the most basic questions here. What is the job description of logical words? It is thought far too often that their task is to serve as guideposts to inference. If so, you could presumably change their meaning by changing the inference rules that govern them.

This view is fundamentally misguided. What *the real job of logical words is*, is *to serve the representation of reality in language.* All communication and all inference relies on such representation. It is in virtue of their representational role that logical words can serve to mediate inferences. And this representational role is not played by means of formal rules of inference. Hence simply to postulate changes in the formal rules of inference or in truth tables does not automatically result in interpretable language. Likewise, changes in the customary truth tables do not automatically have a semantical interpretation. Postulating an indefinite truth value might have some plausibility in the case of a notion like baldness which admits of differences in degree. But in other cases an attribution of an indefinite truth value has no meaning whatsoever. I know what it is for an object to be yellow and likewise what it is for it not to be yellow. But if you tell me that your shirt is neither yellow nor not-yellow, I can associate a meaning to your utterance only by assuming that you are using language in a nonliteral sense. Likewise, there is no limbo between existence and non-existence.

The only way out seems to change our concept of negation from the contradictory negation to something else. This means, for instance, in one of our examples assuming that there are separate criteria of deciding whether it is yellow and of deciding whether it is not-yellow. Naturally, this latter option would mean using negation in some way different from the familiar contradictory negation.

The bald man paradox is not a good example here because in it we are tempted (or at least willing) to think of baldness and hirsuteness as separate concepts. Then the truth value gaps would be simply the cases intermediate between these two independently identifiable concepts. But if they are separate concepts, the use of negation in (P.1)* – (P.2)* and (C)* is not correct.

More generally, one cannot simply postulate truth value gaps or postulate unconventional truth tables and expect to obtain an interpretable language. Language is holistic in that different logical constants do not do their representational job in isolation but always in cooperation with each other and with other notions. Physicists have apparently had more semantical common sense than philosophers here. They have been seriously puzzled by problems like Schrödinger's cat but they have not followed Reichenbach's advice and tried simply to stuff the poor beast into a truth value gap.

I agree that the postulation of truth value gaps is a step in the right direction. However, it does not suffice alone to clarify the problem situation. Rather, the introduction of truth value gaps is like the discovery of a brilliant new gambit in chess that nevertheless does not work unless the overall rules of the game are changed.

But what is the counterpart in such a change of rules in the case of the game of logic? The postulation of truth value gaps means that atomic sentences with suitable nonlogical predicates are neither true nor false; in other words, that the law of excluded middle holds for them. But you cannot change your logic only in its application to some particular subject matter. Truth value gaps for atomic sentences force you to acknowledge truth value gaps for other propositions, including quantified ones, thus making the interpretation of the language even more difficult. What would the reality have to be like in order to allow us to make sense of a world where mammals neither exist nor do not exist?

Hence the philosophers who have postulated truth value gaps ought not to have used the usual textbook logic. *If we change our logic as it applies to atomic propositions, we must change it across the board.* Thus one of the things that the bald man tells us is that it simply is a mistake to give up the law of excluded middle for some predicates but otherwise use the conventional first-order logic which presupposes the law. But logicians defending truth value gaps have never explained what an alternative logic might be and how it is to be interpreted. Yet only such an interpretation could make clear sense of even our familiar logical truths. For instance, if vague predicates are allowed, the following sentences will no longer remain logically true even though they must be taken to be logical truths according to our usual logic employed in the usual attempts to solve the sorites paradox:

(1) There is someone such that if he or she can solve this problem, anyone can.

(2) Someone will commit suicide if he fails in business if and only if someone will commit suicide if everyone fails in business. (Peirce's paradox, see Hintikka 2006.)

These apparently have the respective logical forms

(1)* $(\exists x)(A(x) \supset (\forall y)(A(y))$

(2)* $(\exists x)(F(x) \supset S(x)) \leftrightarrow (\exists x)((\forall y)F(y) \supset S(x))$

Yet (1)* – (2)* cannot be correct representations of the logical forms of (1) – (2), for (1)* – (2)* are logical truths of the usual first-order logic while (1) – (2) are obviously not. Hence it is inappropriate in dealing with the truth value gaps idea to employ the received first-order logic. As (1) – (2) illustrate, it does not even yield right results in all cases. Another reason is that when logic is formalized, it becomes very awkward to maintain a distinction between laws that hold for atomic formulas and others. For atomic formulas are typically used in formalizations as placeholders for arbitrary formulas. Yet, in the logic used in the usual discussions of paradoxes, no explanation is given of how truth value gaps of complex predicates involving quantifiers could be interpreted.

The basic insight that is not heeded in the earlier discussion is, thus, that *if we admit truth value gaps we must have in our logic a strong dual negation* that does not obey the law of excluded middle. This is perhaps the most important thing that the bald man can tell us. Otherwise, we cannot deal, for instance, with the totality of truth value gaps nor with the meaning of any third (indefinite) truth value.

Fortunately, there exists a logic which can serve these purposes. This logic is known as independence-friendly (IF) logic. (For it, see e.g. Hintikka 1996 and 2004.) It is more fundamental than our ordinary first-order logic, which could be called, somewhat inaccurately historically, the Frege-Russell logic. As has been repeatedly pointed out, Frege-Russell logic is unnecessarily restricted in its expressive power. There are several crucial logical and mathematical ideas that cannot be expressed in the ordinary first-order logic but can be expressed by means of IF logic, including equicardinality, infinity and topological continuity. Now we can add to the services IF logic can perform the solution of the sorites paradox.

The source of this additional representational power is purely logical. Part of the semantical (representative) function of quantifiers is to express the dependence and independence of variables. A bound variable y depends (in the interpreted real life sense) on another variable x if and only if it is bound to a quantifier (Q_2y) that depends formally on the quantifier (Q_1x) to which x is bound. Now, in Frege-Russell languages such formal dependence is expressed by the nesting of scopes. But the scope relation is of a very special kind, being transitive and antisymmetric, among other features. Hence it cannot serve to express patterns of dependence relations that do not all have all these features. Hence, it cannot do its whole job.

This restriction is removed in IF logic, which thus can be formulated notationally by allowing a quantifier (Q_2y) to be independent of another one, say (Q_1x), even though it occurs in its (formal) scope. This is indicated by writing it (Q_2y/Q_1x). (For IF logic, see e.g. Hintikka 1996, 1998 and 2002(b).) A semantics is thus obtained for IF logic from the usual game-theoretical semantics for ordinary first-order logic by letting every move prompted by an independent quantifier (Q_2y/Q_1x) to be informationally independent of the move prompted by (Q_1x). The truth of a sentence S is defined in the usual way as the existence of a winning strategy for the verifier in the corresponding semantical game G(S). The falsity of S is defined as the existence of a winning strategy for the falsifier.

It turns out that even when all the rules for semantical games are otherwise the usual ("classical") ones, some games are not determinate: Neither the verifier nor the falsifier has a winning strategy in G(S). The S is neither true nor false. The law of excluded middle does not hold in IF logic.

This conclusion is perfectly straightforward. There are nevertheless several remarkable things about it. First, IF logic ensues inevitably from purely classical semantics (classical rules for semantical games). The only novelty is the admission of independence. This feature is needed for IF logic to have its required expressive power. IF logic is thus not a "nonclassical" logic. It should rather be called neoclassical or perhaps hyperclassical logic.

A further point is especially important here. Not only does IF logic lead us to acknowledge sentences that are neither true nor false. Let's call them indefinite. The game-theoretical semantics for IF logic shows what it means for a sentence S to be indefinite. The truth of S in a model ("world") M means the existence of a naming strategy for the verifier. This is a definite combinational (set-theoretical) fact about M. The same holds for the falsity of S. By the same token, the fact that S is indefinite means that M has certain objective combinatory features. If I know that S is indefinite, I know that the world has those features. Hence *game-theoretical semantics assigns an objective meaning to all attributions of a truth value, including the indefinite one.*

At this point, a look at the wider context of the discussion here is in order. The possibility of making interpretational sense of attributions of indefinite truth value does not only clarify the meaning of formal logics without the law of excluded middle. It opens new avenues of the applications of logical methods. For instance, why is the axiomatic method so useful? In it we study a class of all possible scenarios which are models of the given system. Instead of having to go out to the world and investigate such possible scenarios experimentally and observationally, an axiomatist can study them by drawing logical consequences from the axioms, without needing more than pencil and paper (and, in these days, a computer). This advantage is even independent of whether the logical consequence relations needed are themselves mechanizable.

Now usually the class of models ("scenarios") so examined are the ones in which the axioms are true. But we could equally well study different classes of models in the same way, for instance, the class of models in which the axioms are not false. In mathematical logic, this strategy has in fact been employed under the title "no-counterexample interpretation". This interpretation has been used in an interesting way by Georg Kreisel (see Feferman 1996). Now we can see its significance in a wider context.

The reader might have wondered about what was said about (1)* – (2)* above. If the only difference between received first-order logic and IF first-order logic lies in allowing independence by means of the slash notation, how can the switch from the former affect the status of formulas like (1)* – (2)*? They do not contain any slashes. The answer is that we have to use IF logic not only when slashes are present but also as soon as the basic nonlogical predicates allow for truth value gaps, that is, violations of the law of excluded middle. This puts in perspective what has been said earlier in this paper. We can now see the precise technical source of the difficulties into which the theorists relying on truth value gaps have run into. Their mistake is not to have realized that the use of IF logic is necessary not only when slashes are present but also when the primitive predicates are assumed to have truth value gaps. They have continued blithely to use the traditional first-order logic.

The logical laws of IF first-order logic differ essentially from those of the received first-order logic only because *tertium non datur* sometimes fails. We therefore have to look out for such failures. What are they like? For one thing, (S ∨ ~S) is no longer always true. The conditional (A ⊃ B) is still equivalent to (~A ∨ B), but what it now says is that if A is true *or indefinite*, B is true.

Now we have in fact reached a solution of the bald man paradox. What has been shown is that the paradox should be discussed in terms of IF logic rather than traditional first-order logic. If so, the inference from (P.1)* and (P.2)* to (C)* is indeed valid. But the second premise is not true. For what it now says is that if a man with n hairs is bald or neither-bald-nor-not-bald, then a man with n+1 hairs is bald. This is obviously not true.

This result is as full a solution to the sorites paradox as one can hope. Not only does it show how to avoid the paradox in our formal discourse. It offers an educated diagnosis of how the paradox comes about. In a sense, it vindicates the truth value gaps idea. For the only substantial assumption we had to make is the failure of the law of excluded middle, which is a generalization of the truth value gaps assumption. And the failure of the law of excluded middle is made inevitable by IF logic, which, in turn, is indispensable for our logical language to serve its task adequately. The evidence for the unavoidability of IF logic is independent of the sorites paradox. In this sense, *we have solved the paradox without any additional assumptions* beyond simply the meaning of logical constants, which is built into the interpretation of the underlying logic. In other words, the paradox has been solved by purely logical means without evoking any pragmatic or other extralogical ideas. The solution can be considered as being merely a consistent way of implementing the initial hunch that the source of the paradox is the existence of "truth value gaps", that is, of cases where the law of excluded middle fails for the crucial predicate. The reason why earlier attempts to carry out this idea led to further problems is that in the absence of IF logic these attempts were not coherent. The *tertium non datur* was not allowed to fail in all cases.

Does this solve the sorites paradox for good? Can editors of philosophy journals now safely reject all new papers on the sorites paradox? Yes and no. The solution reached here is definitive, but it is not the whole story. It seems to concern only languages whose whole logic is captured by IF logic. But in natural language and, indeed, in every adequately strong language the contradictory negation is also present. The discovery of IF logic shows a remarkable thing about these languages (see Hintikka 2002a). Since the classical semantical rules that are obviously followed in our native logic give rise to a negation different from the contradictory one, it must be present in any normal language. Hence, *in all sufficiently strong languages there are inevitably two different negations present,* the strong (dual, game-theoretical) negation and the contradictory negation. This is the case also in natural languages even if there is usually only one kind of negation lexically and syntactically present. This insight throws, in fact, an interesting light on several phenomena in the grammar of negation in natural languages. (For such grammar, see e.g. Horn 1989.)

Thus our examination of the sorites paradox needs a second stage because one can try to formulate the paradox by means of a logic that goes beyond IF logic. For one thing, we have to consider (and use) richer languages in which the two negations are present. Why does the sorites paradox not arise in them? Or does it?

The answer to this question obviously depends on a preliminary one. How can the contradictory negation be introduced into a purely IF language? Since the classical semantical rules give rise to the strong negation, we can, without changing our semantics, introduce the contradictory negation, ¬S (as distinguished from the strong negation ~S), only by stipulating that ¬S is true if and only if S is *not* true. Here the stress indicates, as it often does in natural language, contradictory negation in the metalanguage. It follows that ¬ can only occur sentence-initially but not prefixed to open formulas.

The resulting logic has been called extended IF logic. It can be considered our true basic logic. It is elementary in the sense that in its interpretation we need no applications of *tertium non datur* to closed totalities and in the sense that in its applications (in the relevant "language games") no infinite moves are needed. Extended IF logic is symmetrical between its two halves. The original IF half is equivalent to the Σ_1^1 fragment of second-order logic while its mirror image half ("the ¬ half") is equivalent to the Π_1^1 fragment. In IF logic there exists a complete disproof procedure but no complete proof procedure, while in the negated half the converse situation prevails. Each half admits of a truth predicate formulated in the same half. Algebraically, the extended IF logic exhibits a well-known structure: it is a Boolean algebra with an operator (see Hintikka 2004).

The question now becomes: Does the sorites paradox rear its head in extended IF logic? In order for it to arise, we would have to use a conditional in (P.2)** whose substitution instances are not of the form

(3)   ~A(n) ∨ A(n+1)

but rather of the form

(4)   ¬A(n) ∨ A(n+1).

(Here n is a free numerical variable.) Similar things can be pointed out *mutatis mutandis* of (P.2)*.

But in (4) the contradictory negation ¬ occurs prefixed to an open formula. (In a negation normal form of the principle of induction, ¬

would occur within the scope of an existential quantifier.) This is not allowed in the extended IF logic. Hence the paradox of the bald man does not arise in it, either.

Here we can, in fact, have a closer view of the sources of the paradox. In the case of the bald man version, it can trivially be assumed that

(5) $\quad \neg A(n+1) \vee A(n)$.

Together with (4), this implies

(6) $\quad \neg A(n) \vee A(n)$.

In other words, the principle of mathematical induction does not produce the paradox unless the predicate (simple or complex) figuring in it obeys the law of excluded middle. This condition is not satisfied in the bald man case. In (P.2)** the (usually complex) operative predicate is $A(x)$ while in (P.2)* it is $(H(x,n) \supset B(x))$, which has truth value gaps as soon as $B(x)$ does.

What makes formulas like (1)* – (2)* paradoxical is the fact that in them ordinary first-order logic with its law of excluded middle is used in the teeth of truth value gaps. The same holds of the usual form of mathematical induction.

We can thus express one important upshot of these observations in general terms by saying that *the principle of mathematical induction is applicable* (without additional assumptions) *only if the law of excluded middle applies to the operative predicate.* Thus the problematic of the sorites paradox is closely connected with the principle of mathematical induction. No wonder that the two have frequently been discussed together.

Because of this close connection, it is now in order to make a third round of comments on negation and on mathematical induction. They lead us to yet new logical languages, languages beyond an IF language extended by sentence-initial contradictory negation. It is, in fact, possible to give a meaning to the contradictory negation $\neg$ also when it is used within the scope of a quantifier. However, doing that requires going beyond extended IF logic and beyond game-theoretical semantics. Consider, for the purpose of seeing this, the following pair of sentences:

(7) $(\forall x)(A(x) \mathbin{\&} \sim(\forall y)A(y))$

(8) $(\forall x)(\neg\sim A(x) \mathbin{\&} \neg(\forall y)A(y))$

Either of them can be taken to represent the logical form of the negation of (1). However, game-theoretical semantics does not assign an interpretation to (8). There nevertheless is an obvious way of doing so. What we can do is to say that (8) is true if and only if for each member m of the domain of individuals, the verifier has a winning strategy in the game connected with

(9) $\neg\sim A(m) \mathbin{\&} \neg(\forall y)A(y).$

In (9), the contradictory negation $\neg$ occurs only sentence-initially, since m is a constant. Hence it can be interpreted game-theoretically. The existence of a winning strategy for the verifier for all values of m can indeed serve as a truth condition of (9). But then (8) can be taken to be true because (9) is true for all values of m. For $\neg\sim A(m)$ says merely that A(m) is not false. It can still fail to be true, making $(\forall x)A(x)$ true. Hence (1) can be false, and hence it is not logically true. This reflects the fact that (9) is true for all values of m only if A(x) satisfies the law of excluded middle.

The extended truth condition illustrated by this example is nonelementary in a way the truth condition for IF logic is not, for it relies heavily on considering the domain of discourse (which may be infinite) as a closed totality. However, if we are not worried about this nonelementary ("infinitary") character of the resulting logic, we can allow (starting from extended IF logic and hence moving entirely on the first-order level) arbitrary nesting of the contradictory negation. The result is a first-order logic that is as strong as the entire received second-order logic. (This is shown in Hintikka 2006.)

In the resulting logic, the original formal derivation of the paradox would be valid. In other words, in the new, much stronger language (unlike the two languages considered earlier, viz. unextended and extended IF languages) there is a valid formalized version of the argument that leads to the paradox. It is (P.1)* − (P.2)* with (A ⊃ B) interpreted as

($\neg$A $\vee$ B). Does that mean that we still have a paradox on our hands? No, because the only thing that can now be concluded is that the formal argument would no longer be a faithful representation of the ordinary language argument from (P.1) – (P.2) to (C). Since in (P.2)* and in (P.2)** the conditional occurs within the scope of a universal quantifier, from the point of view of traditional first-order logic their semantics is that of second-order logic. The validity of the formal derivation of the paradox belongs to the same category of puzzles as the validity of the well known first-order formulas that seem to express ordinary language sentences that are not valid intuitively at all.

We have encountered cases in point in the form of (1)* – (2)*. Their status as logical truths turns on the interpreting of (A $\supset$ B) as ($\neg$A $\vee$ B), which makes (1)* – (2)* wrong representations as the logical form of the ordinary language statements (1) – (2) when truth value gaps are present, as they implicitly are in (1) – (2). Now we have seen that from the point of view of the fully extended IF logic the pardoxicality of the bald man paradox is of the same kind as that of (1)* – (2)*.

In terms of the reconstruction of second-order logic on the first-order level, we can thus give an interpretation to the principle of mathematical induction where even the second premise (P.2)** is understood as having the form

(10) $(\forall n)(\neg A(n) \vee A(n+1))$.

However, the principle of mathematical induction is still on this construal applicable only when the predicate A(x) obeys the law of excluded middle, that is, does not have any truth value gaps. For if it does, then it could be false, in that for some n, A(n) might be neither true nor false (and hence make $\neg$A(n) and (10) true) while A(n+1) is false. Thus the role of mathematical induction in the resolution of the paradox remains the same.

The conclusions we have reached concerning mathematical induction might not at first sight affect the uses of mathematical induction in mathematics, for the nonlogical predicates and functions used there do not usually admit truth value gaps. But a complex predicate may fail to obey the law of excluded middle also because of the presence of informational independencies (expressed by the slash). Now, IF logic with its slash notation has not been explicitly used in ordinary mathematical research. However, it has been pointed out repeatedly (see e.g. Hodges 2006) that informational independence and hence IF logic has tacitly been used in actual mathematical reasoning. Hence one can be sure of the applicability of mathematical induction to some complicated special case only by first analyzing the propositions involved logically so as to ascertain that no hidden informational independencies lurk somewhere in them. Otherwise an actual working mathematician might end up as a victim of the sorites paradox. Whether such fallacies are actually found in the mathematical literature requires a wider search of possible examples and a closer examination of them than can be undertaken here. The sorites paradox is hence not merely philosophical logicians' self-inflicted problem. The bald man calls our attention to a significant issue in the actual proof methods of mathematics.

What has been found here has also more philosophical implications than can be explored in a single paper. For instance, the fact that mathematical induction does not work for concepts with truth value gaps provides a perspective on the thought of those philosophers who have emphasized the unsharpness of most of our actual concepts, for instance, Wittgenstein with his "family resemblance" idea. Such philosophers can be expected to have doubts about mathematical induction, too.

In particular, it has been claimed that the presence of unsharp concepts in our language means that it does not have a definite logical structure, a structure that can be captured by an explicit formalization. This claim has been refuted here. We have even seen what the main vehicle of the formalization of unsharp concepts is: IF logic. This puts the entire theory of vagueness into a new light. Is the bald man perhaps telling us that IF logic is, among other things, the true "fuzzy logic"?

References

Feferman, Solomon (1996). "Kreisel's 'unwinding' program". In *Kreiserliana*, ed. By Piergiorgio Odifreddi. Wellesley, MA: A.K. Peters, pp. 247-273 (especially sec. 2).

Fine, Kit (1975). "Vagueness, truth, and logic". *Synthese* 30, pp. 265-300.

Hanfling, Oswald (2000). "What is wrong with sorites arguments?" *Analysis* 61, pp. 29-35.

Hintikka, Jaakko (1996). *The Principles of Mathematics Revisited*. Cambridge: Cambridge University Press.
Hintikka, Jaakko (1998). "Truth definitions, Skolem functions and axiomatic set theory". *Bulletin of Symbolic Logic* 4, pp. 303-337.
Hintikka, Jaakko (2002a). "Negation in logic and in natural language". *Linguistics and Philosophy* 25, pp. 585-600.
Hintikka, Jaakko (2002b). "Hyperclassical logic (aka IF logic) and its implications for logical theory". *Bulletin of Symbolic Logic* 8, pp. 404-423.
Hintikka, Jaakko (2004). "What is the true algebra of logic?" In *First-Order Logic Revisited*, ed. by Vincent Hendricks et al. Berlin: Logos Verlag, pp. 117-128.
Hintikka, Jaakko (2006). "Truth, negation and some basic concepts of logic". In *The Age of Alternative Logics,* ed. by Johan van Bentham et al.. Dordrecht: Springer, pp. 195-220.
Hodges, Wilfred (2006). "The logic of quantifiers". In *The Philosophy of Jaakko Hintikka* (Library of Living Philosophers), ed. by Randall Auxier. Chicago: Open Court, pp. 521-534.
Horgan, Terry (1990). "Psychologistic semantics, robust vagueness, and the philosophy of language". In *Belief Systems in Language: Studies in Linguistic Prototypes,* ed. by S.L. Tsohatzidis. London: Routledge, pp. 535-557.
Horn, Laurence R. (1989). *A Natural History of Negation*. Chicago: The University of Chicago Press.
Keefe, Rosanna, and Peter Smith, eds. (1987). *Vagueness: A Reader*. Cambridge, MA.: MIT Press.
Lakoff, George (1975). "Hedges: A study in meaning criteria and the logic of fuzzy concepts". In *Contemporary Research in Philosophical Logic and Linguistic Semantics,* ed. by D. Hockney et al. Dordrecht: D. Reidel, pp. 221-272.
Sanford, David (1975). "Borderline logic". *American Philosophical Quarterly,* pp. 29-39.
Sainsbury, R.M. (1995). *Paradoxes*, second ed. Cambridge: Cambridge University Press.
Tye, Michael (1989). "Supervaluationism and the law of excluded middle". *Analysis* 49, pp. 141-143.
Tye, Michael (1990). "Vague objects". *Mind* 99, pp. 535-558.
Tye, Michael (1994). "Sorites paradoxes and the semantics of vagueness". *Philosophical Perspectives* 8, pp. 189-206.

2

Tonk - A Full Mathematical Solution

ARNON AVRON

1 Introduction: The Tonk Problem

There is a long tradition (see e.g. Hodges 2001, Sundholm 2002) starting from Gentzen 1969, according to which the meaning of a connective is determined by the introduction and elimination rules which are associated with it. The supporters of this thesis usually have in mind natural deduction systems of a certain ideal type (explained in Section 3 below). Unfortunately, the handling of classical negation already requires rules which are not of that type. This problem can be solved in the framework of multiple-conclusion Gentzen-type systems (also first introduced in Gentzen 1969),where the elimination rules are replaced by left introduction rules.

The thesis according to which the meaning of a connective is given by its introduction (and elimination) rules was strongly challenged by Prior in "The Runabout Inference Ticket" (1960). In that paper he introduced his famous "connective" Tonk (denoted below by T). This "connective" has two rules of the ideal type. The introduction rule allows inference of $\varphi T\psi$ from φ. The elimination rule allows inference of ψ from $\varphi T\psi$. In the presence of Tonk every formula can be derived from any other

Hues of Philosophy.
Anat Biletzki (ed.).
Copyright © 2010.

formula, making the "logic" which is "defined" by any system which includes this "connective" trivial.

Prior's paper has made it clear that not every combination of "ideal" introduction and elimination rules can be used for defining a connective. Some constraints should be imposed on the set of rules. Such a constraint was indeed suggested by Belnap in his famous "Tonk, Plonk and Plink" (1962): $\Diamond$ should be *conservative*, in the sense that if $T \vdash \varphi$ is derivable using them, and $\Diamond$ does not occur in $T \cup \{\varphi\}$, then $T \vdash \varphi$ can also be derived without using the rules for $\Diamond$. This solution to the Tonk problem has at least two problematic aspects:

1. Belnap did not provide any effective necessary and sufficient criterion for checking whether a given set of rules is conservative in the above sense. Without such a criterion every connective defined by inference rules (without an independent denotational semantics) is suspected of being a Tonk-like connective, and should not be used until a proof is given that it is "innocent".

2. Belnap formulated the condition of conservativity only with respect to the basic deduction framework, in which no connectives are assumed. But nothing in what he wrote excludes the possibility of a system G having two connectives, each of them "defined" by a set of rules which is conservative over the basic system B, while G itself is not conservative over B. If this happens then it will follow from Belnap's thesis that each of the two connectives is well-defined and meaningful, but they cannot exist together. Such a situation is no less paradoxical than that described by Prior. In order to prevent it one should demand a much stronger conservativity condition than that actually given by Belnap, and it might not even be clear how to *formulate* this stronger condition. Demanding conservativity of the rules for $\Diamond$ over *any* system not involving $\Diamond$ seems too strong, and, in any case, totally useless: how can one prove, for example, that even the standard rules for conjunction are conservative over *any* system not involving conjunction? (Indeed they are *not* always conservative — and there are plenty of examples of this phenomenon in relevance logics and other logics.)

Later attempts of solutions of the Tonk problem insisted on closer connections between the introduction and the elimination rules for a given connective than those implicit in Belnap's condition of conservativity. Usually, it is demanded that the introduction and elimination rules should precisely match (see e.g. Hodges 2001, Sundholm 2002), in the sense that the elimination rules should be derived from the introduction rules by some syntactic procedure. In the words of Hodges (2001, following Prawitz and others): "...the elimination rules only allow us to infer from a formula what we had to know in order to introduce the formula". Others (e.g. Bonnay and Simmenauer 2005) emphasize *invertible* rules (either introduction rules or elimination rules) as those defining meaningful connectives. The discussion is still very much alive, but what is common to almost all the works on the subject is a purely syntactic approach to the question what sets of rules define the meaning of connectives. Needless to say, such works may eliminate the immediate threat of the Tonk argument, but cannot persuade someone who does not understand in what sense can rules define the meaning of connectives.[1]

In this paper we suggest a completely different solution for the Tonk problem which will settle it (so we believe) once and for all. Our approach is to translate the debatable philosophical thesis of Gentzen (which Prior tried to refute) into a precise, unquestionable *mathematical theorem*. Our results can (imprecisely) be summarized as follows: *Every acceptable set of rules of the ideal type[2] determines the meanings of the connectives involved, where the meaning is given in terms of effective conventional denotational semantics (that generalizes in a straightforward way the usual semantics of the classical and intuitionistic logics).* To show this we need to first give precise mathematical definitions of the relevant concepts. This is done in sections 2–4. Then we introduce our algebraic semantic framework and formulate the connection between rules and corresponding semantics in terms of general soundness and completeness theorems. The main idea behind our semantics is to allow variable degrees of *non-determinism* in assigning truth values to formulas. This idea makes it possible to develop denotational semantics for rules in a modular way.

Two comments before we turn to the technical part:

---

[1] Some recent work on the Tonk problem can be found in Bonnay and Simmenauer (2005), Cook (2005), Tennant (2005) and Wansing (2006).
[2] The technical term "canonical rule" is used below for "rule of the ideal type".

1. We treat here two frameworks for rules: natural deduction systems, and Gentzen-type systems. For the latter we again consider two types: multiple-conclusion (of which the well-known system for classical logic is paradigmatic) and single-conclusion (of which the well-known system for intuitionistic logic is paradigmatic). Now the problem of Tonk is usually considered in the framework of natural deduction, but the framework of single-conclusion Gentzen-type systems is completely equivalent, and its use allows for greater uniformity in treating multiple-conclusion logics and single-conclusion ones. Hence the emphasis below is on Gentzen-type systems, but our results can easily be translated into the natural deduction framework.

2. The results concerning the multiple-conclusion frameworks have already been presented in Avron and Lev (2001, 2005). In contrast, the treatment of the single-conclusion framework is new.[3]

2 What is a Propositional Logic?

In what follows $\mathcal{L}$ is a *countable* [4] propositional language, $\mathcal{F}$ is its set of wffs, p,q,r denote atomic formulas, ψ,φ,θ denote arbitrary formulas (of $\mathcal{L}$), T,S denote sets of formulas, and Γ,Δ,Σ,Π denote finite sets of formulas. $Fv(X)$ denotes the set of atomic formulas occurring in X. We assume that the atomic formulas of $\mathcal{L}$ are $p_1, p_2, \ldots$ (in particular: $\{p_1, p_2, \ldots, p_n\}$ are the first n atomic formulas of $\mathcal{L}$).

DEFINITION 1.

1. A *Tarskian consequence relation* (*tcr* for short) for $\mathcal{L}$ is a binary relation $\vdash$ between sets of formulas of $\mathcal{L}$ and formulas of $\mathcal{L}$ that satisfies the following conditions:

   strong reflexivity: if $\varphi \in T$ then $T \vdash \varphi$.
   monotonicity: if $T \vdash \varphi$ and $T \subseteq T'$ then $T' \vdash \varphi$.
   transitivity (cut):   if $T \vdash \psi$ and $T,\psi \vdash \varphi$ then $T \vdash \varphi$.

---

[3] A part of it can by now be found in Avron and Lahav (2009), which includes also results about the validity of the cut-elimination theorem for our various systems. Again in the multiple-conclusion case such results have already been proved in Avron and Lev (2001, 2005).

[4] This requirement is not essential, but it is convenient.

2. A *Scott* (Scott 1974a, 1974b) *consequence relation* (*scr* for short) for $\mathcal{L}$ is a binary relation ⊢ between sets of formulas of $\mathcal{L}$ that satisfies the following conditions:

    strong reflexivity: *if* $T \cap S \neq \emptyset$ *then* $T \vdash S$.
    monotonicity: *if* $T \vdash S$ *and* $T \subseteq T'$ *then* $T' \vdash S'$.
    transitivity (cut):   *if* $T \vdash \psi, S$ *and* $T, \psi \vdash S$ *then* $T \vdash S$.

Conventions: Let ⊢ be a tcr. Below $\Gamma \vdash \{\varphi\}$ means that $\Gamma \vdash \varphi$. In particular: $\vdash \{\varphi\}$ (or $\vdash \varphi$) means $\emptyset \vdash \varphi$. On the other hand $\Gamma \vdash$ (or $\Gamma \vdash \emptyset$) means that $\Gamma \vdash \varphi$ for every $\varphi \in \mathcal{F}$. Provided that the cardinality of Δ is at most 1, these conventions will allow us to use statements of the form $\Gamma \vdash \Delta$ even in case ⊢ is a tcr. As usual, we shall write Γ, φ instead of $\Gamma \cup \{\varphi\}$, and Γ, Γ' instead of $\Gamma \cup \Gamma'$.

DEFINITION 2. An scr or tcr ⊢ for $\mathcal{L}$ is *structural* (or substitution-invariant) if for every uniform $\mathcal{L}$-substitution σ and every Γ and Δ, if $\Gamma \vdash \Delta$ then $\sigma(\Gamma) \vdash \sigma(\Delta)$. ⊢ is *finitary* if the following condition holds for all $T, S \subseteq \mathcal{F}$: if $T \vdash S$ then there exist finite $\Gamma \subseteq T$ and $\Delta \subseteq S$ such that $\Gamma \vdash \Delta$. ⊢ is *consistent* (or *non-trivial*) if there exist non-empty Γ and Δ s.t. $\Gamma \nvdash \Delta$.

The following proposition has been proved in Avron and Lev (2005):

PROPOSITION 3. *There are exactly four inconsistent finitary scrs and two inconsistent finitary tcrs in any given language.*

The four inconsistent finitary scrs and the two inconsistent finitary tcrs are trivial. We exclude them therefore from our definition of a *logic*:

DEFINITION 4. A single-conclusion (or Tarskian) propositional *logic* is a pair $\langle \mathcal{L}, \vdash \rangle$, where $\mathcal{L}$ is a propositional language and ⊢ is a tcr for $\mathcal{L}$ which is structural and consistent. Similarly, a multiple-conclusion (or Scottian) propositional *logic* is a pair $\langle \mathcal{L}, \vdash \rangle$, where $\mathcal{L}$ is a propositional language and ⊢ is an scr for $\mathcal{L}$ which is structural and consistent. In both cases $\langle \mathcal{L}, \vdash \rangle$ is finitary if ⊢ is finitary.

3 What Is a (Canonical) Rule?

An examination of the standard examples of rules used in natural deduction systems or Gentzen-type system shows that an ideal logical rule should be an introduction rule or an elimination rule for one connective (where in the Gentzen-type case by an "elimination rule" we mean an introduction rule in the antecedent), and its formulation should include exactly one occurrence of that connective, while no other occurrences of that connective or any other connective should be mentioned anywhere else in it. Moreover: the rule should be *pure* in the sense of Avron (1991) (i.e., there should be no side conditions limiting its application), and its active formulas should be immediate subformulas of its principal formula. The definitions in this section formulate this idea in exact terms.

3.1 The Multiple-conclusion Case

DEFINITION 5. A *sequent* is an expression of the form $\Gamma \Rightarrow \Delta$ where Γ and Δ are finite sets of formulas. A sequent is called a *clause* if it consists of atomic formulas only.

DEFINITION 6. (Avron and Lev 2005)

1. A *canonical rule* is an expression of the form $\{\Pi_i \Rightarrow \Sigma_i\}_{1 \leq i \leq m}/C$, where $m \geq 0$, C is either $\Diamond(p_1, p_2, \ldots, p_n) \Rightarrow$ or $\Rightarrow \Diamond(p_1, p_2, \ldots, p_n)$ for some connective $\Diamond$ of arity n, and for all $1 \leq i \leq m$, $\Pi_i \Rightarrow \Sigma_i$ is a nonempty clause such that $\Pi_i, \Sigma_i \subseteq \{p_1, p_2, \ldots, p_n\}$.[5]

2. An *application* of a canonical rule $\{\Pi_i \Rightarrow \Sigma_i\}_{1 \leq i \leq m}/\Diamond(p_1, \ldots, p_n) \Rightarrow$ is any inference step of the form:

$$\frac{\{\Gamma, \Pi_i^* \Rightarrow \Delta, \Sigma_i^*\}_{1 \leq i \leq m}}{\Gamma, \Diamond(\psi_1, \ldots, \psi_n) \Rightarrow \Delta}$$

where Π_i^* and Σ_i^* are obtained from Π_i and Σ_i (respectively) by substituting ψ_j for p_j (for all $1 \leq j \leq n$), and Γ, Δ are any finite sets of formulas. An application of a canonical rule with a conclusion of the form $\Rightarrow \Diamond(p_1, \ldots, p_n)$ is defined similarly.

Note: While sequents are written in a metalanguage for $\mathcal{L}$ (which includes the extra symbol $\Rightarrow$), a canonical rule is formulated in a meta-metalanguage of $\mathcal{L}$ (which includes another extra symbol: $/$).

---

[5]Recall that $p_1, p_2, \ldots, p_n$ are the first n atomic formulas.

3.2 The Single-conclusion Case

DEFINITION 7. A *positive Horn clause* is a sequent of the form $\Pi \Rightarrow \{q\}$, where q is an atomic formula, and Π is a finite set of atomic formulas. (we shall usually write $\Pi \Rightarrow q$ instead of $\Pi \Rightarrow \{q\}$). A *negative Horn clause* is a sequent of the form $\Pi \Rightarrow$, where Π is a finite set of atomic formulas. A *Horn clause* is either a positive Horn clause or a negative one.

DEFINITION 8. A *single-conclusion sequent* is an expression $\Gamma \Rightarrow \varphi$ where Γ is finite set of formulas, and φ is a formula.

DEFINITION 9.

1. A *canonical (logical) introduction 1-rule* is an expression of the form $\{\Pi_i \Rightarrow \Sigma_i\}_{1 \le i \le m} / \Rightarrow \Diamond(p_1, p_2, \ldots, p_n)$, where $m \ge 0$, $\Diamond$ is a connective of arity n, and for all $1 \le i \le m$, $\Pi_i \Rightarrow \Sigma_i$ is a positive Horn clause such that $\Pi_i \cup \Sigma_i \subseteq \{p_1, p_2, \ldots, p_n\}$.

2. A *canonical (logical) elimination 1-rule* is an expression of the form $\{\Pi_i \Rightarrow \Sigma_i\}_{1 \le i \le m} / \Diamond(p_1, p_2, \ldots, p_n) \Rightarrow$, where $m \ge 0$, $\Diamond$ is a connective of arity n, and for all $1 \le i \le m$, $\Pi_i \Rightarrow \Sigma_i$ is a Horn clause (either positive or negative) such that $\Pi_i \cup \Sigma_i \subseteq \{p_1, p_2, \ldots, p_n\}$.

3. An *application* of the 1-rule $\{\Pi_i \Rightarrow \Sigma_i\}_{1 \le i \le m} / \Diamond(p_1, p_2, \ldots, p_n) \Rightarrow$ is any inference step of the form:

$$\frac{\{\Gamma, \Pi_i^* \Rightarrow \varphi_i\}_{1 \le i \le m}}{\Gamma \Rightarrow \Diamond(\psi_1, \ldots, \psi_n)}$$

where Γ is a finite set of formulas, and for all $1 \le i \le m$, $\Pi_i^* \Rightarrow \varphi_i$ is obtained from $\Pi_i \Rightarrow \Sigma_i$ by substituting ψ_j for p_j ($1 \le j \le n$).

4. An *application* of the 1-rule $\{\Pi_i \Rightarrow \Sigma_i\}_{1 \le i \le m} / \Diamond(p_1, p_2, \ldots, p_n) \Rightarrow$ is any inference step of the form:

$$\frac{\{\Gamma, \Pi_i^* \Rightarrow \varphi_i\}_{1 \le i \le m}}{\Gamma, \Diamond(\psi_1, \ldots, \psi_n) \Rightarrow \theta}$$

where θ is a formula, Γ is a finite set of formulas, and for all $1 \le i \le m$: Π_i^* is obtained from Π_i by substituting ψ_j for p_j

($1 \leq j \leq n$), and $\varphi_i = \psi_j$ in case $\Sigma_i = \{p_j\}$, $\varphi_i = \theta$ in case Σ_i is empty.

Note. We formulated the definition above in terms of Gentzen-type systems. However, we could have formulated them in terms of natural deduction systems instead. The definition of an application of an introduction rule is defined in this context exactly as above, while an application of an elimination 1-rule of the form $\{\Pi_i \Rightarrow \Sigma_i\}_{1 \leq i \leq m} / \Diamond(p_1, p_2, \ldots, p_n) \Rightarrow$ is in the context of natural deduction any inference step of the form:

$$\frac{\{\Gamma, \Pi_i^* \Rightarrow \varphi_i\}_{1 \leq i \leq m+1}}{\Gamma \Rightarrow \theta}$$

where Γ, θ, Π_i^*, φ_i ($1 \leq i \leq m$) are as above, while Π_{m+1}^* is empty, and φ_{m+1} is $\Diamond(\psi_1, \ldots, \psi_n)$.

3.3 Examples

Canonical Rules for Conjunction

The two usual rules for conjunction are: $\{p_1, p_2 \Rightarrow\} / p_1 \wedge p_2 \Rightarrow$ and $\{\Rightarrow p_1, \Rightarrow p_2\} / \Rightarrow p_1 \wedge p_2$ (denoted by $(\wedge \Rightarrow)$ and $(\Rightarrow \wedge)$, respectively). In the multiple-conclusion context applications of these rules have the form:

$$\frac{\Gamma, \psi, \varphi \Rightarrow \Delta}{\Gamma, \psi \wedge \varphi \Rightarrow \Delta} \qquad \frac{\Gamma \Rightarrow \Delta, \psi \quad \Gamma \Rightarrow \Delta, \varphi}{\Gamma \Rightarrow \Delta, \psi \wedge \varphi}$$

In the single-conclusion context applications of these rules have the form:

$$\frac{\Gamma, \psi, \varphi \Rightarrow \theta}{\Gamma, \psi \wedge \varphi \Rightarrow \theta} \qquad \frac{\Gamma \Rightarrow \psi \quad \Gamma \Rightarrow \varphi}{\Gamma \Rightarrow \psi \wedge \varphi}$$

In natural deduction systems applications of these rules have the form:

$$\frac{\Gamma, \psi, \varphi \Rightarrow \theta \quad \Gamma \Rightarrow \psi \wedge \varphi}{\Gamma \Rightarrow \theta} \qquad \frac{\Gamma \Rightarrow \psi \quad \Gamma \Rightarrow \varphi}{\Gamma \Rightarrow \psi \wedge \varphi}$$

Note that the above elimination rule can easily be shown to be equivalent to the combination of the two more usual elimination rules for conjunction. Moreover: this is the form obtained when one applies in the case of conjunction the standard general method for deriving the elimination rule for a connective from the introduction rules for that connective.

Canonical Rules for Disjunction

The two usual classical rules for disjunction are: $\{p_1 \Rightarrow, p_2 \Rightarrow\} / p_1 \vee p_2 \Rightarrow$ and $\{\Rightarrow p_1, p_2\} / \Rightarrow p_1 \vee p_2$ (denoted by $(\vee \Rightarrow)$ and $(\Rightarrow \vee)$, respectively).

In the multiple-conclusion context applications of these rules have the form:

$$\frac{\Gamma, \psi \Rightarrow \Delta \quad \Gamma, \varphi \Rightarrow \Delta}{\Gamma, \psi \vee \varphi \Rightarrow \Delta} \qquad \frac{\Gamma \Rightarrow \Delta, \psi, \varphi}{\Gamma \Rightarrow \Delta, \psi \vee \varphi}$$

In the single-conclusion context applications of the first rule have the form:

$$\frac{\Gamma, \psi \Rightarrow \theta \quad \Gamma, \varphi \Rightarrow \theta}{\Gamma, \psi \vee \varphi \Rightarrow \theta}$$

In the natural-deduction version applications of the same rule have the form:

$$\frac{\Gamma, \psi \Rightarrow \theta \quad \Gamma, \varphi \Rightarrow \theta \quad \Gamma \Rightarrow \psi \vee \varphi}{\Gamma \Rightarrow \theta}$$

In contrast, the second rule is not applicable in the single-conclusion context, since its premise is not a Horn clause. Instead, the following two introduction rules are usually employed in this context: $\{\Rightarrow p_1\} / \Rightarrow p_1 \vee p_2$ and $\{\Rightarrow p_2\} / \Rightarrow p_1 \vee p_2$. Applications of these rules then have the form:

$$\frac{\Gamma \Rightarrow \psi}{\Gamma \Rightarrow \psi \vee \varphi} \qquad \frac{\Gamma \Rightarrow \varphi}{\Gamma \Rightarrow \psi \vee \varphi}$$

It should be noted that this splitting of the introduction rule for $\vee$ (on the right hand side of a sequent) can be done in the multiple-conclusion case as well. Similarly, the introduction for $\wedge$ on the left hand side of a sequent can be split into two simpler rules by using the same method.

Canonical Rules for Implication

The two usual rules for implication are: $\{\Rightarrow p_1, p_2 \Rightarrow\} / p_1 \supset p_2 \Rightarrow$ and $\{p_1 \Rightarrow p_2\} / \Rightarrow p_1 \supset p_2$ (denoted by $(\supset\Rightarrow)$ and $(\Rightarrow\supset)$, respectively).

In the multiple-conclusion context applications of these rules have the form:

$$\frac{\Gamma \Rightarrow \Delta, \psi \quad \Gamma, \varphi \Rightarrow \Delta}{\Gamma, \psi \supset \varphi \Rightarrow \Delta} \qquad \frac{\Gamma, \psi \Rightarrow \Delta, \varphi}{\Gamma \Rightarrow \Delta, \psi \supset \varphi}$$

In the single-conclusion context applications of these rules have the form:

$$\frac{\Gamma \Rightarrow \psi \quad \Gamma, \varphi \Rightarrow \theta}{\Gamma, \psi \supset \varphi \Rightarrow \theta} \qquad \frac{\Gamma, \psi \Rightarrow \varphi}{\Gamma \Rightarrow \psi \supset \varphi}$$

In the natural-deduction version applications of the first rule have the form:

$$\frac{\Gamma \Rightarrow \psi \quad \Gamma, \varphi \Rightarrow \theta \quad \Gamma \Rightarrow \psi \supset \varphi}{\Gamma \Rightarrow \theta}$$

Again this form of the rule is obviously equivalent to the more usual one (from $\Gamma \Rightarrow \psi$ and $\Gamma \Rightarrow \psi \supset \varphi$ infer $\Gamma \Rightarrow \varphi$).

Canonical Rules for Semi-Implication

Suppose we introduce a "semi-implication"[6] $\rightsquigarrow$ with the following two rules: $\{\Rightarrow p_1, p_2 \Rightarrow\} / p_1 \rightsquigarrow p_2 \Rightarrow$ and $\{\Rightarrow p_2\} / \Rightarrow p_1 \rightsquigarrow p_2$.

---

[6]This connective was independently introduced in Gurevich and Neeman (2009) for completely different purposes.

In the multiple-conclusion context applications of these rules have the form:

$$\frac{\Gamma \Rightarrow \Delta, \psi \quad \Gamma, \varphi \Rightarrow \Delta}{\Gamma, \psi \rightsquigarrow \varphi \Rightarrow \Delta} \qquad \frac{\Gamma \Rightarrow \Delta, \varphi}{\Gamma \Rightarrow \Delta, \psi \rightsquigarrow \varphi}$$

In the single-conclusion context applications of these rules have the form:

$$\frac{\Gamma \Rightarrow \psi \quad \Gamma, \varphi \Rightarrow \theta}{\Gamma, \psi \rightsquigarrow \varphi \Rightarrow \theta} \qquad \frac{\Gamma \Rightarrow \varphi}{\Gamma \Rightarrow \psi \rightsquigarrow \varphi}$$

Again in the natural-deduction version applications of the first rule are equivalent to MP for $\rightsquigarrow$ (from $\Gamma \Rightarrow \psi$ and $\Gamma \Rightarrow \psi \rightsquigarrow \varphi$ infer $\Gamma \Rightarrow \varphi$).

Canonical Rules for Negation

The two usual classical rules for negation are: $\{\Rightarrow p_1\}/\neg p_1 \Rightarrow$ and $\{p_1 \Rightarrow\}/\Rightarrow \neg p_1$ (denoted by $(\neg \Rightarrow)$ and $(\Rightarrow \neg)$, respectively).

In the multiple-conclusion case applications of these rules have the form:

$$\frac{\Gamma \Rightarrow \Delta, \psi}{\Gamma, \neg \psi \Rightarrow \Delta} \qquad \frac{\Gamma, \psi \Rightarrow \Delta}{\Gamma \Rightarrow \Delta, \neg \psi}$$

In the single-conclusion context applications of the first rule have the form:

$$\frac{\Gamma \Rightarrow \psi}{\Gamma, \neg \psi \Rightarrow \theta}$$

In the natural-deduction version applications of the same rule have the form:

$$\frac{\Gamma \Rightarrow \psi \quad \Gamma \Rightarrow \neg \psi}{\Gamma \Rightarrow \theta}$$

In contrast, the second rule is not applicable in the single-conclusion context, since its premise is not a positive Horn clause. There is indeed no satisfactory introduction rule for negation in the single-conclusion context.[7]

Canonical Rules for "Tonk"

Prior's two rules for Tonk are: $\{p_2 \Rightarrow\} / p_1 T p_2 \Rightarrow$ and $\{\Rightarrow p_1\} / \Rightarrow p_1 T p_2$ (denoted by $(T \Rightarrow)$ and $(\Rightarrow T)$, respectively).

In the multiple-conclusion context applications of these rules have the form:

$$\frac{\psi, \Gamma \Rightarrow \Delta}{\varphi T \psi, \Gamma \Rightarrow \Delta} \qquad \frac{\Gamma \Rightarrow \Delta, \varphi}{\Gamma \Rightarrow \Delta, \varphi T \psi}$$

In the single-conclusion context applications of these rules have the form:

$$\frac{\psi, \Gamma \Rightarrow \theta}{\varphi T \psi, \Gamma \Rightarrow \theta} \qquad \frac{\Gamma \Rightarrow \varphi}{\Gamma \Rightarrow \varphi T \psi}$$

In the natural-deduction version applications of the first rule have the form:

$$\frac{\psi, \Gamma \Rightarrow \theta \quad \Gamma \Rightarrow \varphi T \psi}{\Gamma \Rightarrow \theta}$$

This form of the rule is obviously equivalent to that given by Prior: from $\Gamma \Rightarrow \varphi T \psi$ infer $\Gamma \Rightarrow \psi$.

4 What Sets of Rules Are Acceptable?

We start by defining canonical systems, and the consequence relations associated with them.

---

[7]There are two standard solutions to this problem. One is to allow (in the Gentzen-type framework) the use of sequents of the form $\Gamma \Rightarrow$ (in addition to normal single-conclusion sequents of the form $\Gamma \Rightarrow \varphi$). This is somewhat artificial in the single-conclusion context. The other solution is to introduce a propositional constant $\bot$, together with the axioms $\bot, \Gamma \Rightarrow \psi$ for every Γ and ψ, and then to define $\neg \psi$ as $\psi \supset \bot$. Note that the axioms for $\bot$ can be viewed as applications of the canonical 1-elimination rule: $\emptyset / \bot \Rightarrow$.

DEFINITION 10.

1. A multiple-conclusion Gentzen-type system is called *canonical* if its axioms are the sequents of the form $\Gamma \Rightarrow \Delta$ where $\Gamma \cap \Delta \neq \emptyset$, Cut is one of its rules, and each of its other rules is a canonical logical rule.

2. A single-conclusion Gentzen-type system (or an n.d. system) is called *canonical* if its axioms are the sequents of the form $\Gamma \Rightarrow \varphi$ where $\varphi \in \Gamma$, Cut is one of its rules,[8] and each of its other rules is either a canonical 1-introduction rule or a canonical 1-elimination rule.

DEFINITION 11. Let G be a canonical Gentzen-type system of either sort.

1. The tcr $\vdash_G^{seq}$ between *sequents* is defined by: $S \vdash_G^{seq} s$ (where s is a sequent and S is a set of sequents) if there is a derivation in G of s from S.

2. The tcr $\vdash_G$ between *formulas* which is induced by G is defined by: $T \vdash_G \varphi$ iff there exists a finite $\Gamma \subseteq T$ such that $\vdash_G^{seq} \Gamma \Rightarrow \varphi$.

3. If G is multiple-conclusion then the scr $\vdash_G^m$ between formulas is defined by: $T \vdash_G^m S$ if there exist finite $\Gamma \subseteq T, \Delta \subseteq S$ such that $\vdash_G^{seq} \Gamma \Rightarrow \Delta$.

PROPOSITION 12.

1. *If G is canonical then $\vdash_G$ is a structural and finitary tcr. If G is also multiple-conclusion then $\vdash_G^m$ too is structural and finitary.*

2. $T \vdash_G \varphi$ iff $\{\Rightarrow \psi \mid \psi \in T\} \vdash_G^{seq} \Rightarrow \varphi$.

The last proposition does not guarantee that every canonical system induces a *logic* (in the sense of Definition 4). For this the system should satisfy one more condition:

DEFINITION 13. A canonical (single-conclusion or multiple-conclusion) Gentzen-type system G is called coherent if it satisfies the following

---

[8]This condition is not necessary for natural deduction systems.

condition: If both $S_1/\Diamond(p_1, p_2,\ldots, p_n) \Rightarrow$ and $S_2/ \Rightarrow \Diamond(p_1, p_2,\ldots, p_n)$ are rules of G, then the set of clauses $S_1 \cup S_2$ is classically inconsistent (and so the empty clause can be derived from it using cuts).

THEOREM 14. *Let G be a canonical Gentzen-type system (either multiple-conclusion or single-conclusion). $\langle \mathcal{L}, \vdash_G \rangle$ is a single-conclusion logic (i.e. $\vdash_G$ is consistent) iff G is coherent.*[9]

Proof. That the coherence of G implies the consistency of $\vdash_G^m$ was shown in Avron and Lev 2005. Obviously, the consistency of $\vdash_G^m$ implies that of $\vdash_G$. This entails the "if" part of the theorem.

For the converse, assume that G is incoherent. This means that G includes two rules $S_1/\Diamond(p_1, p_2,\ldots, p_n) \Rightarrow$ and $S_2/ \Rightarrow \Diamond(p_1, p_2,\ldots, p_n)$, such that the set of clauses $S_1 \cup S_2$ is classically satisfiable. Let v be an assignment in $\{t, f\}$ that satisfies all the clauses in $S_1 \cup S_2$.

Case 1. Suppose that each element of $S_1 \cup S_2$ is positive. This implies that $p_1,\ldots, p_n, \Delta \Rightarrow q$ is an axiom of G for every $(\Delta \Rightarrow q) \in S_1 \cup S_2$. By applying $S_1/\Diamond(p_1, p_2,\ldots, p_n) \Rightarrow$ and $S_2/ \Rightarrow \Diamond(p_1, p_2,\ldots, p_n)$ to these sequents we get proofs in G of $p_1,\ldots, p_n \Rightarrow \Diamond(p_1,\ldots, p_n)$ and of $p_1,\ldots, p_n, \Diamond(p_1,\ldots, p_n) \Rightarrow r$, where r is an atomic formula different from $p_1,\ldots, p_n$. It follows that $\Gamma \Rightarrow r$ is derivable in G, where $\Gamma = \{p_1,\ldots, p_n\}$ and $r \notin \Gamma$.

Case 2. Suppose that $S_1 \cup S_2$ includes a negative clause. Then v should assign f to at least one atomic formula $r \in \{p_1,\ldots, p_n\}$. Define a formula assignment σ by:

$$\sigma(p) = \begin{cases} r & v(p) = f \\ p & v(p) = t \end{cases}$$

Let $\Gamma = \{p \in \{p_1,\ldots, p_n\} \mid v(p) = t\}$. Then $\vdash_G \Gamma, \sigma(\Delta) \Rightarrow \sigma(q)$ in case $(\Delta \Rightarrow q) \in S_1 \cup S_2$. This is trivial in case $v(q) = t$, since in this case $q \in \Gamma$ and $\sigma(q) = q$. On the other hand, if $v(q) = f$ then $v(p) = f$ for some $p \in \Delta$ (since v satisfies the clause $\Delta \Rightarrow q$). Therefore in this case $\sigma(p) = \sigma(q) = r$, and so again

---

[9] Similarly, if G is multiple-conclusion then $\langle \mathcal{L}, \vdash_G^m \rangle$ is a multiple-conclusion logic iff G is coherent.

$\Gamma, \sigma(\Delta) \Rightarrow \sigma(q)$ is an axiom. We can similarly prove that $\vdash_G \Gamma, \sigma(\Delta) \Rightarrow r$ in case $(\Delta \Rightarrow) \in S_1 \cup S_2$. Now by applying $S_1/\Diamond(p_1, p_2, \ldots, p_n) \Rightarrow$ and $S_2/ \Rightarrow \Diamond(p_1, p_2, \ldots, p_n)$ to these provable sequents we get proofs in G of $\Gamma \Rightarrow \Diamond(p_1, \ldots, p_n)$ and of $\Gamma, \Diamond(p_1, \ldots, p_n) \Rightarrow r$. It follows that $\vdash_G \Gamma \Rightarrow r$, where Γ is a set of atoms such that $r \notin \Gamma$ (because $v(r) = f$).

In both cases we have found a set of atoms Γ and atom r not in Γ such that $\vdash_G \Gamma \Rightarrow r$. This easily entails that $\vdash_G p \Rightarrow r$ for some atomic $p \neq r$. Hence $\vdash_G$ is not consistent.

Note. The last theorem implies that coherence is a minimal demand from any acceptable canonical system. As will be shown in the next section, this minimal condition is in fact sufficient for the existence of an effective denotational semantics. Hence it is not necessary to demand that the introduction and elimination rules of a canonical system should exactly match, or that one of these two sets would be derivable from the other. All that is needed is that there will be no conflict between any two rules of the system.

Examples.

- *Any* subset of the set of rules described above for the connectives $\wedge, \vee, \supset, \rightsquigarrow$ and $\neg$ induces a coherent canonical system.[10] It follows that each such subset R induces a single-conclusion logic. Note that if R does not include $(\Rightarrow \neg)$ (which can be used only in calculi employing multiple-conclusion sequents) then R induces in fact *two* single-conclusion logics (depending on whether the corresponding canonical system G is taken as a calculus of multiple-conclusion sequents or single-conclusion sequents). These two logics are in general different. Note also that R may include no rule at all for some of the connectives, or only one rule, or just two

---

[10] For example: for the first two rules for conjunction described in section 3.3 we have $S_1 = \{p_1, p_2 \Rightarrow\}$, $S_2 = \{\Rightarrow p_1, \Rightarrow p_2\}$, and $S_1 \cup S_2$ is the classically inconsistent set $\{p_1, p_2 \Rightarrow, \Rightarrow p_1, \Rightarrow p_2\}$ (from which the empty sequent can be derived using two cuts). If the first of these two rules $(\wedge \Rightarrow)$ is split into $\{p_1 \Rightarrow\}/p_1 \wedge p_2 \Rightarrow$ and $\{p_2 \Rightarrow\}/p_1 \wedge p_2 \Rightarrow$, then each of these new rules again forms with $(\Rightarrow \wedge)$ a coherent pair. Thus in the case of the first $S_1 \cup S_2 = \{p_1 \Rightarrow, \Rightarrow p_1, \Rightarrow p_2\}$, from which the empty sequent can be derived using a single cut.

rules which do *not* perfectly match, like the rules of $\leadsto$. Still, in all these cases R induces a legitimate logic.
- If the rules of a canonical system G include both of the rules for T ("tonk") described in section 3.3, then G is not coherent: The union of the sets of premises of these two rules is $\{p_2 \Rightarrow, \Rightarrow p_1\}$, and this is a classically consistent set of clauses. It follows that such a system G *does not define a logic* (and this is "what is wrong with tonk").

5 The Semantics Induced by Canonical Systems

In this section we describe the denotational semantics induced by coherent canonical systems.

5.1   The Multiple-conclusion Case

DEFINITION 15. An $\mathcal{L}$-*valuation* is a function $v : \mathcal{F} \to \{t, f\}$,

DEFINITION 16. An $\mathcal{L}$-valuation v is a model of a formula φ if $v(\varphi) = t$. It is a model of a theory $\mathcal{T}$ if it is a model of every $\varphi \in \mathcal{T}$.

Definition 15 by itself imposes no restrictions on valuations (in particular, the truth-values a valuation assigns to a complex formula need not be a function of the truth-values it assigns to atomic formulas). The role of rules is to limit the set of allowed valuations by imposing some constraints on them. The idea is that given a canonical system G, only those valuations which respect all the rules of G should be taken into account.

DEFINITION 17. Let v be an $\mathcal{L}$-*valuation*, and let σ assign a formula in $\mathcal{F}$ to every atomic formula. We say that σ satisfies in v a clause $\Pi \Rightarrow \Sigma$ if $v(\sigma(p)) = f$ for some $p \in \Pi$, or $v(\sigma(q)) = t$ for some $q \in \Sigma$.

DEFINITION 18. Let v be an $\mathcal{L}$-*valuation*.

1. v respects a rule of the form $S/\Rightarrow \Diamond(p_1,\ldots,p_n)$ if $v(\Diamond(\psi_1,\ldots,\psi_n)) = t$ whenever all elements of S are satisfied in v by an assignment σ such that $\sigma(p_i) = \psi_i$ $(1 \leq i \leq n)$.[11]

2. v respects a rule of the form $S/\Diamond(p_1,\ldots,p_n) \Rightarrow$ if $v(\Diamond(\psi_1,\ldots,\psi_n)) = f$ whenever all elements of S are satisfied in v by an assignment σ such that $\sigma(p_i) = \psi_i$ $(1 \leq i \leq n)$.

3. Let G be a canonical Gentzen-type system for $\mathcal{L}$. v is G-legal if it respects all the rules of G.

Examples.

- A valuation v respects the rule $(\supset \Rightarrow)$ iff $v(\varphi \supset \psi) = f$ whenever $v(\varphi) = t$ and $v(\psi) = f$. It respects the rule $(\Rightarrow \supset)$ iff $v(\varphi \supset \psi) = t$ whenever $v(\varphi) = f$ or $v(\psi) = t$. Note that these two constraints are independent of each other (but do not contradict each other). Together the two rules dictate that v treats $\supset$ according to the classical truth-table for $\supset$.

- A valuation v respects the rule $(\leadsto \Rightarrow)$ iff $v(\varphi \leadsto \psi) = f$ whenever $v(\varphi) = t$ and $v(\psi) = f$. It respects the rule $(\Rightarrow \leadsto)$ iff $v(\varphi \leadsto \psi) = t$ whenever $v(\psi) = t$. Note that none of these two rules imposes any constraint in case $v(\varphi) = f$ and $v(\psi) = f$. In other words, valuations which respect both rules treat the semi-implication $\leadsto$ according to the following *non-deterministic matrix* (Avron and Lev 2005, Avron 2005a, Avron 2005b):

$\leadsto$	t	f
t	t	f
f	t	$\{t,f\}$

- A valuation v respects $(T \Rightarrow)$ if $v(\varphi T \psi) = f$ whenever $v(\psi) = f$. It respects $(\Rightarrow T)$ if $v(\varphi T \psi) = t$ whenever $v(\varphi) = t$. The two constraints contradict each other in case both $v(\varphi) = t$ and $v(\psi) = f$. This is a semantic explanation for why the "connective" T is meaningless.

---

[11] The values of $\sigma(q)$ for $q \notin \{p_1,\ldots,p_n\}$ are immaterial here.

THEOREM 19. (Avron and Lev 2001, 2005) *Every canonical multiple-conclusion Gentzen-type system G for $\mathcal{L}$ is strongly sound and complete with respect to the semantics of G-legal valuations. In other words: $T \vdash_G \varphi$ iff every G-legal model of T is also a model of φ.*

5.2 The Single-conclusion Case

DEFINITION 20. A generalized $\mathcal{L}$-*frame* is a triple $\mathcal{W} = \langle W, \leq, v \rangle$ s. t.:

1. $\langle W, \leq \rangle$ is a nonempty partially ordered set.

2. v is a function from $\mathcal{F}$ to the set of persistent functions from W into $\{t, f\}$ (A function $h: W \to \{t, f\}$ is *persistent* if $h(a) = t$ implies that $h(b) = t$ for every $b \in W$ such that $a \leq b$).

Notation: We shall usually write $v(a, \varphi)$ instead of $v(\varphi)(a)$.

DEFINITION 21. A generalized $\mathcal{L}$-frame $\langle W, \leq, v \rangle$ is a model of a formula φ if $v(\varphi) = \lambda a \in W.t$ (i.e.: $v(a, \varphi) = t$ for every $a \in W$). It is a model of a theory T if it is a model of every $\varphi \in T$.

DEFINITION 22. Let $\langle W, \leq, v \rangle$ be a generalized $\mathcal{L}$-frame, let $a \in W$, and let σ assign a formula in $\mathcal{F}$ to every atomic formula. We say that σ satisfies in a a positive Horn clause $\Pi \Rightarrow q$ if for every $b \geq a$, either $v(b, \sigma(p)) = f$ for some $p \in \Pi$, or $v(b, \sigma(q)) = t$. We say that σ satisfies in a a negative Horn clause $\Pi \Rightarrow$ if $v(a, \sigma(p)) = f$ for some $p \in \Pi$.

Note: Because of the persistence condition, σ satisfies in a a positive Horn clause of the form $\Rightarrow q$ iff $v(a, \sigma(q)) = t$.

DEFINITION 23. Let $\mathcal{W} = \langle W, \leq, v \rangle$ be a generalized $\mathcal{L}$-frame.

1. Let r be a 1-introduction rule for the n-ary connective $\Diamond$. $\mathcal{W}$ *respects* r if $v(a, \Diamond(\psi_1, \ldots, \psi_n)) = t$ whenever all the premises of r are satisfied in a by an assignment σ such that $\sigma(p_i) = \psi_i$ ($1 \leq i \leq n$).[12][13]

2. Let r be a 1-elimination rule for the n-ary connective $\Diamond$. $\mathcal{W}$ *respects* r if $v(a, \Diamond(\psi_1, \ldots, \psi_n)) = f$ whenever there exists $b \geq a$ in which all

---

[12] Again, the values of $\sigma(q)$ for $q \notin \{p_1, \ldots, p_n\}$ are immaterial here.
[13] Note that since the premises of a 1-introduction rule are all positive, this is equivalent to: for every $b \geq a$, all the premises of R are satisfied in b by σ.

the premises of r are satisfied by an assignment σ such that $\sigma(p_i) = \psi_i$ ($1 \leq i \leq n$).

3. Let G be a canonical Gentzen-type system for $\mathcal{L}$. $\mathcal{W}$ is G-legal if it respects all the rules of G.

Examples.

- A generalized $\mathcal{L}$-frame $\mathcal{W} = \langle W, \leq, v \rangle$ respects the rule ($\supset \Rightarrow$) iff for every $a \in W$, $v(a, \varphi \supset \psi) = f$ whenever there exists $b \geq a$ such that $v(b, \varphi) = t$ and $v(b, \psi) = f$. $\mathcal{W}$ respects ($\Rightarrow \supset$) iff for every $a \in W$, $v(a, \varphi \supset \psi) = t$ whenever for every $b \geq a$, either $v(b, \varphi) = f$ or $v(b, \psi) = t$. Hence the two rules together impose exactly the well-known Kripke semantics for intuitionistic implication (Kripke 1965).

- A generalized $\mathcal{L}$-frame $\mathcal{W} = \langle W, \leq, v \rangle$ respects the rule ($\leadsto \Rightarrow$) iff for every $a \in W$, $v(a, \varphi \leadsto \psi) = f$ whenever there exists $b \geq a$ such that $v(b, \varphi) = t$ and $v(b, \psi) = f$. $\mathcal{W}$ respects ($\Rightarrow \leadsto$) iff for every $a \in W$, $v(a, \varphi \leadsto \psi) = t$ whenever $v(a, \psi) = t$ (recall that this is equivalent to: $v(b, \psi) = t$ for every $b \geq a$). Note that in this case too the two rules for $\leadsto$ do not always determine the value assigned to $\varphi \leadsto \psi$: if $v(a, \psi) = f$, and also $v(b, \varphi) = f$ for every $b \geq a$, then $v(a, \varphi \leadsto \psi)$ is free to be either t or f. So the semantics of this connective is non-deterministic in the single-conclusion context also.

- A valuation v respects ($T \Rightarrow$) if $v(a, \varphi T \psi) = f$ whenever $v(a, \varphi) = f$. It respects ($\Rightarrow T$) if $v(a, \varphi T \psi) = t$ whenever $v(a, \psi) = t$. Again, the two constraints contradict each other in case both $v(a, \varphi) = f$ and $v(a, \psi) = t$. This is a semantic explanation for why the "connective" T is meaningless even in the single-conclusion context.

THEOREM 24. Every canonical single-conclusion system G is strongly sound and complete with respect to the semantics of G-legal generalized frames (i.e.: $\mathcal{T} \vdash_G \varphi$ iff every G-legal model of $\mathcal{T}$ is also a model of φ).

Proof. To prove soundness, assume that $\mathcal{T} \vdash_G \varphi$. Then there is $\Gamma \subseteq \mathcal{T}$ such that $\vdash_G \Gamma \Rightarrow \varphi$. Hence it suffices to show that if $\vdash_G \Gamma \Rightarrow \varphi$, and $\mathcal{W} = \langle W, \leq, v \rangle$ is G-legal, then $\Gamma \Rightarrow \varphi$ has the following property:

For every $a \in W$: (*) either $v(a,\psi) = f$ for some $\psi \in \Gamma$, or $v(a,\varphi) = t$.

Since the axioms of G trivially have this property, and the cut rule obviously preserves it, it suffices to show that the property is preserved also by applications of the logical (canonical) rules of G.

- Suppose $\Gamma \Rightarrow \Diamond(\psi_1,\ldots,\psi_n)$ is derived from $\{\Gamma,\Pi_i^* \Rightarrow \varphi_i\}_{1\leq i\leq m}$ using the rule $\{\Pi_i \Rightarrow q_i\}_{1\leq i\leq m} / \Rightarrow \Diamond(p_1,p_2,\ldots,p_n)$, and assume that all the premises of this application have the required property. We show that so does its conclusion. Let $a \in W$. If $v(a,\psi) = f$ for some $\psi \in \Gamma$, then obviously (*) holds for a and $\Gamma \Rightarrow \Diamond(\psi_1,\ldots,\psi_n)$. Assume otherwise. Then the persistence condition implies that $v(b,\psi) = t$ for every $\psi \in \Gamma$ and $b \geq a$. Hence our assumption concerning $\{\Gamma,\Pi_i^* \Rightarrow \varphi_i\}_{1\leq i\leq m}$ entails that for every $b \geq a$ and $1 \leq i \leq m$, either $v(b,\psi) = f$ for some $\psi \in \Pi_i^*$, or $v(b,\varphi_i) = t$. It follows that for $1 \leq i \leq m$, $\Pi_i \Rightarrow q_i$ is satisfied in a by an assignment σ such that $\sigma(p_j) = \psi_j$ $(1 \leq j \leq n)$. Since $\mathcal{W}$ respects the rule $\{\Pi_i \Rightarrow q_i\}_{1\leq i\leq m} / \Rightarrow \Diamond(p_1,p_2,\ldots,p_n)$, this implies that $v(a,\Diamond(\psi_1,\ldots,\psi_n)) = t$.

- Suppose $\Gamma,\Diamond(\psi_1,\ldots,\psi_n) \Rightarrow \theta$ is derived from $\{\Gamma,\Pi_i^* \Rightarrow \varphi_i\}_{1\leq i\leq m}$ using the rule $\{\Pi_i \Rightarrow \Sigma_i\}_{1\leq i\leq m} / \Diamond(p_1,p_2,\ldots,p_n) \Rightarrow$, and assume that all the premises of this application have the required property. We show that so does its conclusion. Let $a \in W$. If $v(a,\psi) = f$ for some $\psi \in \Gamma$ or $v(a,\theta) = t$, then we are done. Assume otherwise. Then $v(a,\theta) = f$, and (by the persistence condition) $v(b,\psi) = t$ for every $\psi \in \Gamma$ and $b \geq a$. Hence again our assumption concerning $\{\Gamma,\Pi_i^* \Rightarrow \varphi_i\}_{1\leq i\leq m}$ entails that for every $b \geq a$ and $1 \leq i \leq m$, either $v(b,\psi) = f$ for some $\psi \in \Pi_i^*$, or $v(b,\varphi_i) = t$. This immediately implies that every positive premise of the rule is satisfied in a by any assignment σ such that $\sigma(p_i) = \psi_i$ $(1 \leq i \leq n)$. Since $v(a,\theta) = f$, it also implies that $v(a,\psi) = f$ for some $\psi \in \Pi_i^*$ whenever the premise $\Pi_i \Rightarrow \Sigma_i$ is negative (i.e. Σ_i is empty and so φ_i is θ). Hence the negative premises of the rule are also satisfied in a by σ. Since $a \geq a$, and $\mathcal{W}$ respects the rule $\{\Pi_i \Rightarrow \Sigma_i\}_{1\leq i\leq m} / \Diamond(p_1,p_2,\ldots,p_n) \Rightarrow$, it follows that $v(a,\Diamond(\psi_1,\ldots,\psi_n)) = f$, as required.

Next we turn to prove completeness. So let G be a canonical Gentzen-type system for $\mathcal{L}$, and assume that $\mathcal{T}_0 \nvdash_G \varphi_0$. We construct a G-legal generalized $\mathcal{L}$-frame $\mathcal{W}$ which is a model of $\mathcal{T}_0$ but not of φ_0.

Given a formula ψ, call a theory $\mathcal{T}$ ψ-*maximal* if $\mathcal{T} \nvdash_G \psi$, but $\mathcal{T}' \vdash_G \psi$ whenever $\mathcal{T}'$ is a proper extension of $\mathcal{T}$. Obviously, if $\mathcal{T} \nvdash_G \psi$, then $\mathcal{T}$ can be extended to a theory $\mathcal{T}^*$ which is ψ-maximal. In particular: $\mathcal{T}_0$ can be extended to a φ_0-maximal theory $\mathcal{T}_0^*$. Now let $\mathcal{W} = \langle W, \subseteq, v \rangle$, where W is the set of all extensions of $\mathcal{T}_0^*$ which are ψ-maximal for some formula ψ, and v is defined as follows:

$$v(\mathcal{T}, \varphi) = \begin{cases} t & \varphi \in \mathcal{T} \\ f & \varphi \notin \mathcal{T} \end{cases}$$

Obviously, $\mathcal{W}$ is a generalized $\mathcal{L}$-frame (note that $\mathcal{T}_0^* \in W$, and so W is not empty). Since $\mathcal{T}_0 \subseteq \mathcal{T}_0^* \subseteq \mathcal{T}$ for every $\mathcal{T} \in W$, the definition of v implies that $\mathcal{W}$ is a model of $\mathcal{T}_0$. Since $v(\mathcal{T}_0^*, \varphi_0) = f$, $\mathcal{W}$ is not a model of φ. Therefore it only remains to prove that $\mathcal{W}$ is G-legal.

Lemma 1 If $\mathcal{T} \in W$, and $\mathcal{T} \vdash_G \varphi$, then $v(\mathcal{T}, \varphi) = t$.

Proof of Lemma 1: Since $\mathcal{T} \in W$, $\mathcal{T}$ is ψ-maximal for some ψ. Now assume that $\varphi \notin \mathcal{T}$. Then $\mathcal{T} \cup \{\varphi\} \vdash_G \psi$. Since $\mathcal{T} \vdash_G \varphi$, this implies $\mathcal{T} \vdash_G \psi$. A contradiction. It follows that $\varphi \in \mathcal{T}$, and so $v(\mathcal{T}, \varphi) = t$.

Lemma 2 Let $\mathcal{T} \in W$, and let $\Delta \cup \{\varphi\}$ be a set of formulas of $\mathcal{L}$. Then $\mathcal{T} \cup \Delta \vdash_G \varphi$ iff the following condition holds for every $\mathcal{T}' \in W$ such that $\mathcal{T} \subseteq \mathcal{T}'$: either $v(\mathcal{T}', \psi) = f$ for some $\psi \in \Delta$, or $v(\mathcal{T}', \varphi) = t$.

Proof of Lemma 2: Assume that $\mathcal{T} \cup \Delta \vdash_G \varphi$. Let $\mathcal{T}'$ be an element of W such that $\mathcal{T} \subseteq \mathcal{T}'$, and $v(\mathcal{T}', \psi) = t$ for every $\psi \in \Delta$. Then $\Delta \subseteq \mathcal{T}'$ by definition of v, and so $\mathcal{T} \cup \Delta \subseteq \mathcal{T}'$. It follows that $\mathcal{T}' \vdash_G \varphi$, and so $v(\mathcal{T}', \varphi) = t$ by Lemma 1.

For the converse, assume that $\mathcal{T} \cup \Delta \nvdash_G \varphi$. Extend $\mathcal{T} \cup \Delta$ to a φ-maximal theory $\mathcal{T}'$. Then $\mathcal{T}' \in W$, $\mathcal{T} \subseteq \mathcal{T}'$, $\Delta \subseteq \mathcal{T}'$ and $\varphi \notin \mathcal{T}'$. Hence the definition of v implies that the condition above fails for $\mathcal{T}'$.

Now we show that $\mathcal{W}$ respects the rules of G.

Assume first that $\{\Pi_i \Rightarrow q_i\}_{1 \leq i \leq m} / \Rightarrow \Diamond(p_1, p_2, \ldots, p_n)$ is an introduction rule of G, and that for every $1 \leq i \leq m$, $\Pi_i \Rightarrow q_i$ is satisfied in $\mathcal{T} \in W$ by an assignment σ such that $\sigma(p_j) = \psi_j$

$(1 \leq j \leq n)$. By Lemma 2 this implies that $T \cup \{\sigma(p) \mid p \in \Pi_i\} \vdash_G \sigma(q_i)$ for $1 \leq i \leq m$. This means that for $1 \leq i \leq m$, $\vdash_G \Gamma_i \cup \{\sigma(p) \mid p \in \Pi_i\} \Rightarrow \sigma(q_i)$ for some finite $\Gamma_i \subseteq T$. Let $\Gamma = \bigcup_{i=1}^{m} \Gamma_i$. Then $\vdash_G \Gamma \cup \{\sigma(p) \mid p \in \Pi_i\} \Rightarrow \sigma(q_i)$ for $1 \leq i \leq m$. By applying the rule $\{\Pi_i \Rightarrow q_i\}_{1 \leq i \leq m} / \Rightarrow \Diamond(p_1, p_2, \ldots, p_n)$ to these sequents we get that $\vdash_G \Gamma \Rightarrow \Diamond(\psi_1, \ldots, \psi_n)$, where $\Gamma \subseteq T$. Hence $T \vdash_G \Diamond(\psi_1, \ldots, \psi_n)$, and so $v(T, \Diamond(\psi_1, \ldots, \psi_n)) = t$ by Lemma 1.

Now assume that $\{\Pi_i \Rightarrow \Sigma_i\}_{1 \leq i \leq m} / \Diamond(p_1, p_2, \ldots, p_n) \Rightarrow$ is an elimination rule of G, that T and T' are elements of W such that $T \subseteq T'$, and that for every $1 \leq i \leq m$, $\Pi_i \Rightarrow \Sigma_i$ is satisfied in T' by an assignment σ such that $\sigma(p_j) = \psi_j$ $(1 \leq j \leq n)$. Let $1 \leq i \leq m$. If $\Sigma_i = \{q_i\}$ then like in the previous case there is a finite $\Gamma_i \subseteq T'$ such that $\vdash_G \Gamma_i \cup \{\sigma(p) \mid p \in \Pi_i\} \Rightarrow \sigma(q_i)$. If $\Sigma_i = \emptyset$ then the satisfaction of $\{\Pi_i \Rightarrow \Sigma_i\}$ in T' means that there exists $q \in \Pi_i$ such that $v(T', \sigma(q)) = f$, and so $\sigma(q) \notin T'$. Now T' is θ-maximal for some θ. It follows that $T' \cup \{\sigma(q)\} \vdash_G \theta$, and so $\vdash_G \Gamma_i \cup \{\sigma(p) \mid p \in \Pi_i\} \Rightarrow \theta$ for some finite $\Gamma_i \subseteq T'$. Let $\Gamma = \bigcup_{i=1}^{m} \Gamma_i$. Then:

- $\vdash_G \Gamma \cup \{\sigma(p) \mid p \in \Pi_i\} \Rightarrow \sigma(q_i)$ in case $\Sigma_i = \{q_i\}$.
- $\vdash_G \Gamma \cup \{\sigma(p) \mid p \in \Pi_i\} \Rightarrow \theta$ in case $\Sigma_i = \emptyset$..

By applying $\{\Pi_i \Rightarrow \Sigma_i\}_{1 \leq i \leq m} / \Diamond(p_1, p_2, \ldots, p_n) \Rightarrow$ to these provable sequents we get $\vdash_G \Gamma, \Diamond(\psi_1, \ldots, \psi_n) \Rightarrow \theta$. Hence $T', \Diamond(\psi_1, \ldots, \psi_n) \vdash_G \theta$. It follows that $\Diamond(\psi_1, \ldots, \psi_n) \notin T'$. Since $T \subseteq T'$, this implies that $\Diamond(\psi_1, \ldots, \psi_n) \notin T$. Hence $v(T, \Diamond(\psi_1, \ldots, \psi_n)) = f$, as required.

6 Applications of the Semantics

In general, in order for a denotational semantics of a propositional logic to be useful and effective, it should be *analytic*. This means that to determine whether a formula φ follows from a theory T, it suffices to consider *partial* valuations, defined on the set of subformulas of formulas in $T \cup \{\varphi\}$. Now we show that the semantics of G-legal framesis analytic in this sense.

Notation: For $\mathcal{W} = \langle W, \leq, v \rangle$ and $a \in W$

- $a \vdash'_\mathcal{W} \Pi \Rightarrow \Sigma$ denotes that $v(a, \varphi) = f$ for some $\varphi \in \Pi$, or $v(a, \varphi) = t$ for some $\varphi \in \Sigma$.

- $a \vDash_{\mathcal{W}}^v \Pi \Rightarrow \Sigma$ denotes *that* $b \vDash_{\mathcal{W}}^v \Pi \Rightarrow \Sigma$ for every $b \geq a$.

DEFINITION 25. Let G be a canonical Gentzen type system for $\mathcal{L}$. A G-legal *semiframe* is a triple $\mathcal{W}' = \langle W, \leq, v' \rangle$ such that:

1. $\langle W, \leq \rangle$ is a nonempty partially ordered set.

2. v' is a partial function from the set of formulas of $\mathcal{L}$ into the set of persistent functions from W into $\{t, f\}$ such that:

   - $\mathcal{F}'$, the domain of v', is closed under subformulas.
   - v' respects the rules of G on $\mathcal{F}'$ (e.g.: if r is an introduction rule for an n-ary connective $\Diamond$, and $\Diamond(\psi_1, \ldots, \psi_n) \in \mathcal{F}'$, then $v(a, \Diamond(\psi_1, \ldots, \psi_n)) = t$ whenever all the premises of r are satisfied in a by an assignment σ such that $\sigma(p_i) = \psi_i$ ($1 \leq i \leq n$)).

THEOREM 26. Let G be a *coherent* canonical Gentzen type system for $\mathcal{L}$. Then the semantics of G-legal frames is *analytic* in the following sense: if $\mathcal{W}' = \langle W, \leq, v' \rangle$ is a G-legal *semiframe*, then v' can be extended to a function v such that $\mathcal{W} = \langle W, \leq, v \rangle$ is a G-legal frame.

Proof. Let $\mathcal{W}' = \langle W, \leq, v' \rangle$ be a G-legal semiframe. We recursively extend v' to a total function v. For atomic p we let $v(p) = v'(p)$ if $v'(p)$ is defined, and $v(p) = \lambda a \in W.t$ (say) otherwise. For $\varphi = \Diamond(\psi_1, \ldots, \psi_n)$ we define $v(\varphi)$ for every $a \in W$ by: $v(\varphi, a) = f$ iff there exist an elimination rule r with $\Diamond(p_1, \ldots, p_n) \Rightarrow$ as its conclusion, and an element $b \geq a$ of W, such that all premises of r are satisfied in b (with respect to $\langle W, \leq, v \rangle$) by an assignment σ such that $\sigma(p_j) = \psi_j$ ($1 \leq j \leq n$). Note that the satisfaction of the premises of r by σ in elements of W depends only on the values assigned by v to $\psi_1, \ldots, \psi_n$, so the recursion works, and v is well defined. Moreover: since $\mathcal{W}' = \langle W, \leq, v' \rangle$ is a G-legal semiframe, it can easily be proved by induction that $v(\varphi) = v'(\varphi)$ when the latter is defined. >From the definition of v it immediately follows also that $v(\varphi)$ is a persistent function for every φ (so $\mathcal{W} = \langle W, \leq, v \rangle$ is a generalized $\mathcal{L}$-frame), and that $\mathcal{W}$ respects all the elimination rules of G. Therefore it only remains to prove that it respects also the introduction rules of G. So let $\{\Pi_i \Rightarrow q_i\}_{1 \leq i \leq m} / \Rightarrow \Diamond(p_1, p_2, \ldots, p_n)$ be such a rule, and assume that $a \vDash_{\mathcal{W}}^v \{\sigma(p) \mid p \in \Pi_i\} \Rightarrow \sigma(q_i)$ ($1 \leq i \leq m$). We should show that $v(a, \Diamond(\psi_1, \ldots, \psi_n)) = t$. Assume otherwise. Then there exists $b \geq a$ and an elimination rule $\{\Delta_j \Rightarrow \Sigma_j\}_{1 \leq j \leq k} / \Diamond(p_1, p_2, \ldots, p_n) \Rightarrow$, such that

$b \vDash_W^t \{\sigma(p) \mid p \in \Delta_j\} \Rightarrow \{\sigma(p) \mid p \in \Sigma_j\}$ for $1 \leq j \leq k$. Since $b \geq a$, our assumption about a implies that $b \vDash_W^t \{\sigma(p) \mid p \in \Pi_i\} \Rightarrow \sigma(q_i)$ for $1 \leq i \leq m$. It follows that by defining $u(p) = v(b, \sigma(p))$ we get a valuation u in $\{t, f\}$ which satisfies all the clauses in the union of $\{\Pi_i \Rightarrow q_i \mid 1 \leq i \leq m\}$ and $\{\Delta_j \Rightarrow \Sigma_j \mid 1 \leq j \leq k\}$. This contradicts the coherence of G.

The following two theorems are now easy consequences of Theorem 26 and the soundness and completeness theorems of the previous section:

THEOREM 27. *Let G be a canonical Gentzen type system. Then G is decidable (and so it is decidable whether $\Gamma \vdash_G \varphi$, where φ is a formula and Γ a finite set of formulas).*

THEOREM 28. *Let G_1 be a coherent canonical Gentzen-type system in a language $\mathcal{L}_1$, and let G_2 be a coherent canonical Gentzen-type system in a language $\mathcal{L}_2$. Assume that $\mathcal{L}_2$ is an extension of $\mathcal{L}_1$ by some set of connectives, and that G_2 is obtained from G_1 by adding to the latter canonical rules for connectives in $\mathcal{L}_2 - \mathcal{L}_1$. Then G_2 is a conservative extension of G_1 (i.e. if all formulas in $T \cup \{\varphi\}$ are in $\mathcal{L}_1$ then $T \vdash_{G_1} \varphi$ iff $T \vdash_{G_2} \varphi$).*

Note. The last theorem shows that a very strong form of Belnap's conservativity criterion is valid for coherent canonical systems. Hence it provides a full answer to the second objection concerning this criterion which was raised in the Introduction. The first one is met, of course, by our coherence criterion for canonical systems, since coherence of a finite set of canonical rules can effectively be checked.

Acknowledgment

This research was supported by THE ISRAEL SCIENCE FOUNDATION (grant No 809-06).

References

Avron, A. (1991). "Simple consequence relations". *Information and Computation,* Vol. 92, pp. 105–139.

Avron, A. (2005). "Non-deterministic Matrices and Modular Semantics of Rules". In *Logica Universalis*, ed. by J.-Y. Beziau. Birkhüser Verlag, pp. 149–167.

Avron, A. (2005). "Logical Non-determinism as a Tool for Logical Modularity: An Introduction". In *We Will Show Them: Essays in Honor of Dov Gabbay*, ed. by S. Artemov, H. Barringer, A. S. d'Avila Garcez, L. C. Lamb, and J. Woods. College Publications, Vol. 1, pp. 105–124.

Avron A. and I. Lev (2001). "Canonical Propositional Gentzen-Type Systems". In *Proceedings of the 1st International Joint Conference on Automated Reasoning (IJCAR 2001)*, ed. by R. Goré, A Leitsch, T. Nipkow. LNAI 2083, Springer Verlag, 529–544.

Avron, A. and I. Lev, (2005). "Non-deterministic Multiple-valued Structures". *Journal of Logic and Computation*, Vol. 15, pp. 241–261.

Avron, A. and O. Lahav (2009). "Canonical Constructive Systems". In *Proceedings of the 18th international Conference on Automated Reasoning with Analytic Tableaux and Related Methods (TABLEAUX 2009)*, ed. by M. Giese, A. Waaler. LNAI 5607, Springer, pp. 62–76.

Belnap, N. D. (1962). "Tonk, Plonk and Plink". *Analysis*, Vol. 22, pp. 130–134.

Bonnay, D. and B. Simmenauer (2005). "Tonk Strikes Back". *Australian Journal of Logic*, Vol. 3, pp. 33–44.

Cook, R. T. (2005). "What's Wrong with Tonk?" *Journal of Philosophical Logic*, Vol. 34, pp. 217–226.

Gurevich Y. and I. Neeman (2009). "The Logic of Infons". *Microsoft Research Tech Report MSR-TR-2009-10*.

Hodges, W. (2001). "Elementary predicate logic". In *Handbook of Philosophical Logic*, ed. by D. M. Gabbay and F. Guenthner, 2nd ed., Vol. 1, pp. 1–129.

Sundholm, G. (2002). "Proof theory and meaning". In *Handbook of Philosophical Logic*, ed. by D. M. Gabbay and F. Guenthner, 2nd ed., Vol 9, pp. 165–198.

Kripke, S. (1965). "Semantical Analysis of Intuitionistic Logic I". In *Formal Systems and Recursive Functions*, ed. by J. Crossly and M. Dummett. Amsterdam: North-Holland, pp. 92–129.

Gentzen, G. (1969). "Investigations into logical deduction". In *The Collected Works of Gerhard Gentzen*, ed. by M. E. Szabo. Amsterdam: North Holland, pp. 68–131.

Prior, A.N. (1960). "The Runabout Inference Ticket". *Analysis*, Vol. 21, pp. 38–39.

Scott, D. S. (1974). "Completeness and axiomatization in many-valued logics". In *Proc. of the Tarski symposium*, volume XXV of *Proc. of Symposia in Pure Mathematics*. American Mathematical Society, pp. 411–435.

Scott, D. S. (1974). "Rules and derived rules". In *Logical theory and semantical analysis*, ed. by S. Stenlund. pages Dordrecht: Reidel, pp. 147–161.

Tennant, N. (2005). "Rule-circularity and the Justification of Deduction". *The Philosophical Quarterly* Vol. 55, pp. 625–648.

Wansing, H. (2006). "Connectives Strangers than Tonk", *Journal of Philosophical Logic* Vol. 35, pp. 653–660.

3

Davidson's Notion of Supervenience

ORON SHAGRIR

1 Introduction

In his "Mental Events", Donald Davidson introduces supervenience into the philosophy of mind, and argues that it is consistent with his anomalous monism. There has been much discussion ever since on whether supervenience can secure dependence and anomalism at the same time. Curiously, however, there has been little effort to explicate what Davidson means by supervenience; philosophers typically assume Kim's conception of supervenience. My aim here is to explicate the passages in which Davidson discusses supervenience. I argue that Davidson's supervenience is very different from the one assumed in contemporary philosophy of mind. I suggest that Davidson's supervenience is not dependence in the sense of some deeper metaphysical relation, but a constraint on the attribution of mental and physical predicates. Whether this notion can be reconciled with anomalism is something I discuss elsewhere.

Ruth Manor stressed the importance of a precise analysis of philosophical concepts, without being dogmatic about former distinctions. Her 2001 article on pragmatics and semantics is a wonderful exemplar of this approach: she recognizes the importance of a distinction between the concepts, but argues that there are essential links between them. I want to

Hues of Philosophy.
Anat Biletzki (ed.).
Copyright © 2010.

believe that I bring Ruth's approach to the analysis of Davidson's notion of supervenience (who is also concerned, of course, with the semantics of natural languages). Analyzing Davidson's conception of supervenience reveals a different philosophical motivation and a very interesting approach to the complex and intricate relations between the mind and the physical world.

2 The Characterization of Supervenience

Davidson characterizes supervenience in several places. In "Mental Events", he writes:

> Supervenience might be taken to mean that there cannot be two events alike in all physical respects but differing in some mental respect, or that an object cannot alter in some mental respects without altering in some physical respects (1970: 214).

The first part of the sentence is a characterization in terms of *indiscernibility*, namely, that "there cannot be two events alike in all physical respects but differing in some mental respect", that is, there cannot be two events that are physically indiscernible but mentally discernible. The second part is another characterization, one which construes supervenience in terms of covariance: "an object cannot alter in some mental respects without altering in some physical respects", that is, mental changes *co-vary* with physical changes.

In later writings, Davidson provides additional covariance definitions, according to which any mental difference between objects must be accompanied by a physical difference. In "Reply to Harry Lewis", he writes:

> The notion of supervenience, as I have used it, is best thought of as a relation between a predicate and a set of predicates in a language: a predicate p is supervenient on a set of predicates S if for every pair of objects such that p is true of one and not of the other there is a predicate in S that is true of one and not of the other. (1985: 242)

And in "Thinking Causes" he makes a similar claim:

> The idea I had in mind is, I think, most economically expressed as follows: a predicate *p* is supervenient on a set of predicates *S* if and only if *p* does not distinguish any entities that cannot be distinguished by *S*. (1993: 4)

On a charitable reading, Davidson's characterizations are all equivalent.[1] Supervenience is a thesis about the relations between properties or characteristics or respects, e.g. mental and physical properties, which Davidson understands as *predicates*. These properties are ascribed to particulars such as events, objects, and entities. To make things more explicit, let us take two sets of properties, R and S. We can think of R as a set of mental predicates, and of S as a set of physical predicates. We would say that R *supervenes* on S just in case the following condition holds:

> For every M of R and for every pair of objects (events, entities) x and y, if for every P of S, Px ↔ Py (i.e., x and y are S-indiscernible), then Mx ↔ My (i.e., x and y are R-indiscernible).

Let us compare this characterization to Kim's notions of supervenience. First, Kim (1984) famously distinguishes between a strong and a weak reading of this condition. On the strong reading, the condition applies to every pair of possible objects x and y, even if they inhabit "different worlds". On the weak reading, it applies to every pair of objects belonging to the same world (any world), but need not apply to objects across worlds.[2] In "Thinking Causes", Davidson says that his version of supervenience is of the weak sort:

---

[1] The first definition, in terms of indiscernibility, is just the contrapositive formulation of the later covariance definitions. Physically indiscernible objects are mentally indiscernible iff mentally discernible objects are physically discernible. And mentally discernible objects are physically discernible iff for every mental predicate M that distinguishes between x and y (e.g., Mx, but ~My) there is a physical predicate P of S that also distinguishes between x and y.

[2] The standard formulation of these two variants is as follows:
R *strongly* supervenes on S just in case for every M of R, for every pair of worlds *v* and *w*, and for every pair of objects x in *v* and y in *w*, if, for every P of S, Px ↔ Py, then Mx ↔ My.
R *weakly* supervenes on S just in case for every M of R, for every world w and for every pair of objects x and y in *w*, if, for every P of S, Px ↔ Py, then Mx ↔ My.

> Kim himself (correctly, I think) finds my version of supervenience very close to his 'weak' supervenience, and as not entailing connecting laws. (1993: 4, n. 4)

Second, Kim demonstrates that under assumptions of closure of S, strong and weak supervenience are equivalent, respectively, to strong and weak entailment P* → M principles, where P* is a maximal S-property.[3] Unlike Kim, Davidson never explicates supervenience in terms of entailment conditionals P* → M.[4] Third, Kim also introduces a notion of global supervenience, which, arguably, fits better with the thesis of externalism.[5] Davidson, we recall, is an outspoken proponent of externalism. But it turns out that global supervenience is an intricate notion that comes with very different versions.[6] Instead of invoking global supervenience, it suffices to use the individual notions but not limit S to monadic, micro, local, or intrinsic properties; it could include causal relations with the physical environment, and even bits of causal history.[7] S could even in-

---

[3] Maximal S-properties are "the strongest consistent properties constructible" in S (Kim 1984: 58), namely, are composed of "simple" properties by Boolean operations and negation. Here are the precise definitions of the entailment conditions:
Strong entailment: For every M of R there is a maximal S-property P* such that for every world w, and for every object x in w, if x has P*, then x has M.
Weak entailment: For every M of R and for every world w there is a maximal S-property P* such that for every object x in w, if x has P*, then x has M.

[4] Whether Davidson thinks that supervenience implies the P* → M conditionals is an interesting question. At one point, in his "Reply to Harry Lewis", it seems that he does, as he states that "supervenience guarantees that the ontology of the subvenient predicates suffices for the supervenient predicate" (1985: 242). But he goes on to say he was "arguing for... the identity of mental events with physical events" (1985: 243), describing that stance as "ontological reduction". This indicates that ontological reduction is no more than the thesis of monism, which in turn explains why Davidson maintains that "it is obvious that ontological reduction does not entail ... nomological reduction" (1985: 242-243), and that if supervenience is a truism then the "truth of monism is a truism" (1985: 244). So there is absolutely no indication that Davidson sees supervenience as implying the P* → M conditionals, though these conditionals are not explicitly repudiated.

[5] I think it was Petrie (1987) who was first to point out the connection between global supervenience and externalism. See also Shagrir (2002).

[6] For the different varieties of global supervenience see Shagrir (2002) and Bennett (2004).

[7] See Davidson (1990a) for a list of the relevant factors in the attribution of mental states.

clude physical properties of remote objects if such properties are indeed relevant to the ascription of mental properties.[8]

Davidson's statement about his notion of supervenience being of the weak variety is puzzling. The problem with weak supervenience is that it does not support dependence. Weak supervenience is consistent with the scenario in which my counterpart and I have exactly the same physical properties, but different mental properties. But then the mental difference is not due to our physical properties, since nothing related to our physical makeup, including past and present causal relations with their environment, differs. It would thus seem that there *are* mental properties that do not depend on physical properties.[9]

Davidson himself rules out such scenarios. He maintains that in counterfactual scenarios like the Twin-Earth and Swampman thought experiments, the mental differences *are* accompanied by physical differences:

> What I take Burge's and Putnam's imagined cases to show (and what I think the Swampman example shows more directly) is that people who are in all relevant physical respects similar (or 'identical' in the necktie sense) can differ in what they mean or think… But of course there is *something* different about them, even in the physical world; their causal histories are different. (Davidson 1987: 32-33)

Weak supervenience, however, lacks the modal force to support these psychophysical dependencies. Why, then, does Davidson invoke weak supervenience? One answer is that mental properties do not strongly supervene on the intrinsic physical properties of agents, as the Twin-Earth and Swampman examples show. The mental only weakly supervenes on

---

[8] We can thus take as the maximal property the object's complete "world-perspective" in terms of S properties. For a precise characterization see Stalnaker (1996) and Sider (1999). Another way of defining these "global" maximal properties is via closure under quantification. When taking P* to be maximal in this global sense, the strong entailment condition is equivalent to strong global supervenience; see Sider (1999).

[9] I assume that "worlds" refer to counterfactual cases. If, deferring to Davidson's aversion to possible worlds, we stipulate that only cases found in our world will be considered, then the conditionals P* → M will support dependence, but will also be counterfactual-supportive, that is, they will be connecting laws.

intrinsic physical properties: two agents of the same "world" that are physically indiscernible are also mentally indiscernible.[10]

However, this answer is deficient. It is true that the mental does not strongly supervene on intrinsic physical properties, but it does strongly supervene on intrinsic *and extrinsic* physical properties. It thus makes more sense to use strong supervenience over intrinsic and extrinsic physical properties, which reflects dependence, rather than using weak supervenience over intrinsic physical properties alone, which does not reflect dependence. Even if weak supervenience does not lead to psychophysical laws that correlate mental properties and intrinsic physical properties, we still need another argument for showing why strong supervenience does not lead to laws that correlate mental properties with intrinsic and extrinsic physical properties. I return to this issue in the last section.

3 The philosophical Role of Supervenience

Davidson does not say much about the philosophical import of supervenience, but it is clear that his views on this are unique.[11] One respect in which they are unusual has to do with the philosophical role of supervenience. Supervenience is widely upheld by nonreductive monists: those who maintain that every mental event *is* a physical event, but deny psychophysical laws. Many who espouse versions of this view worry that it is insufficiently "materialistic". The concern is that nonreductive monism says too little about the relations between mental and physical *properties*. Although it denies that mental properties are physical properties, it imposes no alternative constraints on the attribution of mental properties. Monism ensures that every object with mental properties also has physical properties, but, beyond that, anything goes: monism is consistent with

---

[10] This line of answer can be extracted from what Davidson says in "Could There Be a Science of Rationality?":
[I]t is only if mental properties are supervenient on the physical properties of the agent that there can be any hope of identifying the mental properties with physical properties, or of finding lawlike connections between the two. If mental properties are supervenient not only on physical properties of the agent but in addition on the physical properties of the world outside the agent, there can be no hope of discovering laws that predict and explain behavior solely on the basis of intrinsic features of agents (1995: 122).
For a detailed discussion of this proposal see Shea (2003) and Shagrir (2009).

[11] Davidson (1994) also takes supervenience to be a central pillar of his philosophy of mind.

the possibility that my physical counterpart has no mentality whatsoever, while my cup of coffee does. Surely a monistic, to say nothing of materialistic, doctrine that allows such wild attributions is worthless. Something must be done to close this gap. And this is where supervenience kicks in. The role of supervenience is to put more significant constraints on the psychophysical relations between mental and physical properties, without reducing mental properties to physical properties. Supervenience, on this account, is a secondary thesis that makes nonreductive monism materialistically kosher.

"Mental Events" gives the impression that supervenience plays this legitimizing role in Davidson's philosophy.[12] After presenting the tenets of anomalous monism, Davidson immediately introduces supervenience, saying that "although the position I describe denies there are psychophysical laws, it is consistent with the view that mental characteristics are in some sense dependent, or supervenient, on physical characteristics" (p. 214). From this we might conclude that Davidson, too, feels obliged to complement his monistic thesis about events with a substantive and positive thesis about the psychophysical relations between predicates. But later on, in his "Reply to Harry Lewis" (1985) and in "Thinking Causes" (1993), it turns out that this is not how Davidson sees the role of supervenience. Declaring that "supervenience in any form implies monism; but it does not imply either definitional or nomological reduction", Davidson reveals that he invoked supervenience to demonstrate that anomalous monism is consistent: "So if (non-reductive) supervenience is consistent (as the syntax-semantics example proves it is) so is *AM* [anomalous monism]" (1993:5).[13]

Contrary to first impressions, then, supervenience is not a secondary thesis intended to correct the deficiencies of the primary doctrine of nonreductive monism. There is no evidence that Davidson deems his ano-

---

[12] This is indeed Kim's impression: "Clearly, mind-body supervenience is a net addition to anomalous monism. By adopting it, Davidson has substantially strengthened his position on the mind-body problem" (2003: 130).

[13] See also "Reply to Harry Lewis" (Davidson 1985: 243-244). Lewis (1985) had objected that supervenience is trivially true, since any two events are discernible in some physical respect. But Davidson is unperturbed by the alleged triviality of supervenience, asserting that "if Lewis is right that it is a truism that there cannot be two events alike in all physical respects, the truth of monism is a truism" (1985: 244), which suggests that the role of supervenience is not to *strengthen* anomalous monism, but to establish its consistency.

malous monism to be in need of reinforcement, whereas we do have evidence that Davidson does not take supervenience to provide such reinforcement. Davidson, of course, does resist the idea that mental properties float freely, as it were, over the physical domain, and does take supervenience as asserting that the mental *depends* on the physical realm. But this claim about dependency is not made as a substantive addition to anomalous monism. Rather, supervenience is used to help establish both monism *and* the consistency of anomalous monism. Davidson deploys supervenience once again in "Thinking Causes", this time to secure the causal relevance of mental properties. The claim made is that supervenience entails that an event's mental properties make a difference to its causal relations.[14]

Davidson's supervenience is also unique with respect to the notion of dependence. Most philosophers, following Kim, maintain that mind-body supervenience is grounded in some deeper *metaphysical* relation. The idea is that any $P^* \to M$ conditional reflects the dependence of M on P^*, and this dependence is a metaphysical determination relation, e.g. identity, constitution, emergence, or realization, which underlies and explains the supervenience relations. It is thus not surprising that, in the context of supervenience, the notions of dependence and determination are often used interchangeably.[15] The implicit assumption is that M depends on P^* by virtue of M's being determined by P^*, whereas determination is understood as a metaphysical determination.[16]

Davidson's notion of dependence is different. The idea that the application of a mental predicate is grounded in some metaphysical determination of the mental by a fixed physical basis is foreign to Davidson's approach. He never hints that the mental depends on the physical by virtue of some metaphysical determination relation; and he certainly does not introduce the more familiar determination relations to substantiate his supervenience thesis.[17] In fact, in "Thinking Causes", the main argument

---

[14] See Davidson (1993: 13-14); the term "properties" is Davidson's.

[15] See, e.g. Bennett (2004).

[16] See Kim, who writes that "mind-body supervenience ... points to the existence of a dependency relation" (1998: 10), which grounds or accounts for it.

[17] Davidson associates supervenience with the term determination only once, in "The Material Mind": "There is a sense in which the physical characteristics of an event (or object or state) *determine* the psychological characteristics; in G.E. Moore's view, psychological

for the causal relevance of mental properties suggests that supervenience is *not* such a determination relation. He asserts: "supervenience as I have defined it does, as we have seen, imply that if two events differ in their psychological properties, they differ in their causal properties (which we assume to be causally efficacious). If supervenience holds, psychological properties make a difference to the causal relations of an event, for they matter to the physical properties, and the physical properties matter to causal relations" (1993: 14). But it is apparent that "make a difference" cannot be understood to mean "determine" in a metaphysical sense. For it refers to the mental-to-physical direction, whereas the pertinent metaphysical relation is in the physical-to-mental direction. It is most unlikely that Davidson would take supervenience to point to metaphysical determination of the mental by the physical, and still claim that supervenience implies that mental properties "matter to the physical properties".

We see that Davidson does not uphold the idea that supervenience reflects metaphysical physical-to-mental determination or a dependence relation. It seems, moreover, that he also rejects the idea that dependence (of mental on the physical) grounds or accounts for supervenience. If anything, it is the other way around. Davidson says that "supervenience gives a sense to the notion of dependence here, enough sense anyway to show that mental properties make a causal difference" (1993: 14). So it is not that dependence accounts for supervenience, but, if anything, dependence is explicated in terms of the supervenience of the mental on the physical.

Lastly, it is telling that Davidson invokes supervenience in *causal* contexts. In discussing the Twin-Earth and Swampman cases, Davidson insists that "of course there is *something* different about them, even in the physical world; their causal histories are different." He later describes supervenience as implying that "mental properties make a causal difference." And he links supervenience with the causal nature of the mental, stating that "Kim, as we noted, thinks my version of supervenience implies that all mental properties could be withdrawn from the world and this would make no difference to causal relations; but this supposition turned out to be incompatible with my understanding of su-

---

concepts are *supervenient* on physical concepts". But he immediately makes clear that determinacy here is nothing more than indiscernibility: "it is impossible for two events (objects, states) to agree in all their physical characteristics ... and to differ in their psychological characteristics" (1973a: 253).

pervenience" (1993: 14); and that "[s]upervenience as I defined it is consistent with ... the assumption that there are no psychophysical laws ... It is not even slightly plausible that there are no important general causal connections between mental and physical properties of events. I have always held that there are such connections" (1993: 14).

Let me sum up the distinctive features of Davidson's supervenience I have mentioned. Supervenience is not a secondary thesis the objective of which is to reinforce anomalous monism. It is not explicated by some deeper metaphysical determination or dependence relation, but, if anything, it is supervenience that gives cogency to the notion of dependence. And it has something to do with the "causal connections between mental and physical properties of events". In addition, he characterizes supervenience in terms of indiscernibility or covariance and not in terms of the entailment $P^* \to M$ conditionals, and declares that his supervenience is of the weak kind. But what are we to make of all this? Can we extract from these remarks a cohesive notion of supervenience? Can we say what supervenience *is*?

4 A Davidsonian Conception of Supervenience

I propose an understanding of supervenience that fits in with Davidson's remarks on supervenience, and with his overall philosophical outlook. Hence even if not exactly Davidson's notion in every detail, this notion of supervenience can be characterized as Davidsonian. The proposal is that supervenience is a thesis about the relations between the application of physical predicates and the procedures of interpretation: the procedures by which an interpreter attributes mental predicates to others on the basis of what they say and do.[18] The thesis is that the relations are not arbitrary, but are constrained by the indiscernibility condition. The constraint is that an attribution of mental difference is always accompanied by some physical difference. Put differently, supervenience states that an interpre-

---

[18] There is an intimate linkage between interpretation and mentality in Davidson's philosophy of mind – between someone's thinking that *p*, and his being interpreted as thinking that *p* on the basis of what he says and does: "Thoughts, desires, and other attitudes are in their nature states we are equipped to interpret; what we could not interpret is not thought" (Davidson 1990b: 88). For a comprehensive analysis of the relations between interpretation and mentality, see Child (1994, chapter 1). See also Davidson (1973b, 1974).

ter ascribes the same mental properties to objects or events unless there is some physical difference about them, i.e. in their physical makeup, environment, or in their causal histories. If there is no such physical difference, there is no mental difference.

But what is the rationale for supervenience? Why impose the no-mental-difference-without-a-physical-difference constraint, if not for metaphysical determination? One answer is that supervenience is a very minimal consistency constraint on psychophysical relations.[19] It does little more than remind us that an interpreter has no reason to ascribe different mental properties to objects with exactly the same physical states, including past and present relations with their environments.[20]

A more interesting answer is that supervenience is an offshoot, at least to some extent, of other constraints. In saying this, I have in mind the linkage between supervenience and the "causal connections between mental and physical properties of events". These causal connections can be characterized in two divergent ways. According to the nomological conception, where there is a causal relation, there is a law, a strict physical law. In particular, causal relations are underwritten by laws that constrain the interpreter to ascribe exactly the same physical properties to two objects that have exactly the same causal interactions with the physical environment. The other characterization is in terms of the crucial role such causal connections play in fixing the content of thoughts, beliefs, and other attitudes. The public and intersubjective character of the mental prompts the interpreter to ascribe the same mental properties to two objects that have exactly the same causal interactions with their environments. Supervenience can be seen as linking the two characterizations by emphasizing that the physical (type) identity that is constrained by laws is accompanied by mental (type) identity that is constrained by intersubjectivity.[21]

---

[19] This was suggested by John Heil in a private communication.

[20] The constraint is also minimal in that it is not very practical, as any two objects differ in some physical respect.

[21] It should be noted, however, that supervenience is *not* the claim that mental states are public, or accessible to the interpreter, viz., that an interpreter can tell, under favorable conditions, what someone thinks and believes. Supervenience is a claim about what the interpreter should do in any given circumstance, namely, assign the same mental properties to objects unless she can identify some physical difference between them. Supervenience does not imply, without further assumptions, that there are favorable circumstances

My proposal, then, is that the supervenience of the mental on the physical is not dependence in the sense of some deeper metaphysical relation. Rather, the mental depends on the physical in the sense that the application of mental predicates is constrained by their supervenience relations with physical predicates, that is, by the indiscernibility condition. But what exactly does this come down to? What is the modal force of this condition?

The supervenience condition should be strong in one sense and weak in another. It should be strong enough to support the counterfactuals in the pertinent thought experiments, where Davidson insists that there is something different in the physical properties of Davidson and Swampman (who differ in their mental states), and something different in the physical properties of Oscar and his Twin-Earth doppelganger. It should be weak enough to justify Davidson's assertion that his notion is of the weak sort, one that does not entail connecting laws.

One way to formulate such a notion is in a two-dimensional system where the "metaphysical" dimension consists of "worlds" that are the usual counterfactual situations, as in the Twin-Earth and Swampman examples.[22] The "epistemic" dimension consists of "scenarios"; these are points-of-view from which we assign extensions to predicates in these worlds. We can think of these scenarios, very roughly, as possible ("epistemic") states along the interpretation process, in which an interpreter correlates mental and physical properties; at each stage she assigns men-

---

under which the physical differences will be accessible to the interpreter. And it does not imply that these physical differences, even if accessible, can serve to tell *what* the mental differences are. The physical differences do not even imply that there is a mental difference, for not every physical difference results in a mental difference. Knowing that there is a mental difference, and knowing what the difference is, calls for knowledge of the conditionals P* → M. But, as I attempt to show below, it does not follow from supervenience that there are any such strong conditionals.

Is supervenience *implied* by the public nature of the mental? I am not sure. The public nature of the mental implies that if two objects differ mentally, this is reflected in some difference with respect to what they say and do. Saying (i.e. meaning) and doing (i.e. acting), however, are not physical properties, but see Heal (1997: 176), who points out that radical interpretation "starts from information, *all* of which is non-semantic" (see also Davidson 1973b: 128). Supervenience could perhaps be seen as the claim that public differences reach all the way to the physical world. Even if the bodily movements and utterances are exactly the same, as in the Twin-Earth and Swampman examples, the attribution of different thoughts must have some *physical* manifestation.

[22] For a two-dimensional approach, see, e.g., Chalmers (2004).

tal properties to objects with certain physical properties. The states differ *not* in the meaning ("secondary intensions") we assign to predicates, but rather in the amount of *evidence* available to the interpreter.

The idea is that supervenience is strong in the metaphysical dimension, but weak in the epistemic dimension. Put more explicitly, we would say that R supervenes on S just in case:

> For every M of R, for every scenario C_i, for every pair of worlds W_j and W_k, viewed from this scenario, and for every pair of objects x in W_j and y in W_k, the following condition holds: if, for every P of S, Px ↔ Py, then Mx ↔ My.

Let me explicate the difference between the dimensions by means of example. It's 10:00 o'clock on Sunday morning. Mary is trying to figure out what Frank wants to do. Two games, one baseball, the other soccer, are scheduled to be broadcast at 11:00 a.m. Mary, being a knowledgeable scientist, knows Frank's physical condition and movements, his physical environment, bits of his physical past, and so on, perhaps even his maximal physical state P*. Based on the evidence she has about his behavior, environment and history, Mary attributes to Frank the mental state M, i.e., a preference to watch soccer over baseball. Supervenience compels Mary to keep the correlation between P* and M across counterfactual situations (worlds along the metaphysical dimension). She will attribute to Dave, who has P*, the mental state M, and she will attribute to Frank's counterpart on Twin-Earth ~M only if he has ~P*.

Supervenience, however, does not compel Mary to keep the correlation between P* and M along the epistemic dimension, across scenarios. Here is what might be a different scenario. Since Mary is no sports fan, she departs to her lab. Returning home at noon, Mary finds Frank on the sofa watching the baseball game. What should she do in this scenario, in light of this new piece of evidence? Mary has no reason to doubt her past attribution of P*, i.e. that Frank had P* at 10:00. The attribution of P* was fixed, via a comprehensive closed system of laws, by the complete physical state before 10:00. Nothing that happened after 10:00 gives her any reason to change this attribution. However, Mary has good reason to revise her previous assessment of Frank's state of mind. In light of the new evidence, she can *now* say that it is better to attribute to individuals with P* M' rather than M. In particular, she can say that although Frank was in a physical state P* at 10:00, he nonetheless had M', i.e., he preferred, at 10 o'clock, to watch baseball over soccer. Having attributed M' to someone with P*, supervenience compels Mary to keep this correlation along the metaphysical dimension. In particular, Mary will attribute to Dave, who has P*, the mental state M', and she will attribute to Frank's counterpart on Twin-Earth ~M', e.g., M, only if he has ~P*.

Can we reconcile this notion of supervenience with the thesis of anomalism? Elsewhere (Shagrir, 2009) I consider and criticize several proposals that appeal to the maximality of physical properties. Another option is taking advantage of the weak aspect of the proposed notion of supervenience. Whether this notion is indeed compatible with anomalism is something I hope to explore in future work.[23]

References

Bennett, Karen (2004). "Global Supervenience and Dependence". *Philosophy and Phenomenological Research* 68, pp. 501–529.

Burge, Tyler (1979). "Individualism and the Mental". In *Midwest Studies in Philosophy: Studies in Metaphysics*, ed. by French, et al. Minneapolis: University of Minnesota Press, pp. 73–122.

Chalmers, David (2004). "Epistemic Two-Dimensional Semantics". *Philosophical Studies*, 118, pp. 153–226.

Child, William (1994). *Causality, Interpretation and the Mind*. Oxford: Oxford University Press.

Davidson, Donald (1970). "Mental Events". In *Experience and Theory*, ed. by L. Foster and J.W. Swanson. Amherst: University of Massachusetts Press, pp. 79-l0l. Reprinted in Davidson 1980.

Davidson, Donald (1973a). "The Material Mind". In *Proceedings of the Fourth International Congress for Logic, Methodology, and Philosophy of Science*, ed. by P. Suppes, L. Henkin, G.C Moisil and A. Joja. Dordrecht: North-Holland, pp. 716–722. Reprinted in Davidson 1980.

Davidson, Donald (1973b). "Radical Interpretation". *Dialectica* 27, pp. 314–328. Reprinted in Davidson 1984.

---

[23] I am grateful to Carl Posy and Ted Sider for invaluable suggestions and discussion. This research was supported by The Israel Science Foundation, grant 857/03-07.

Davidson, Donald (1974). "Belief and the Basis of Meaning". *Synthese* 27, pp. 309–323. Reprinted in Davidson 1984.
Davidson, Donald (1975). "Thought and Talk". In *Mind and Language*, ed. by S. Guttenplan. Oxford: Oxford University Press, pp. 7–23. Reprinted in Davidson 1984.
Davidson, Donald (1980). *Essays on Actions and Events*. Oxford: Clarendon Press.
Davidson, Donald (1983). "A Coherence Theory of Truth and Knowledge". In *Kant oder Hegel?* ed. by D. Henrich. Stuttgart: Klett-Cotta. Reprinted in Davidson 2001.
Davidson, Donald (1984). *Inquiries into Truth and Interpretation*. Oxford: Clarendon Press.
Davidson, Donald (1985). "Reply to Harry Lewis". In *Essays on Davidson: Actions and Events*, ed. by B. Vermazen and M. Hintikka. Oxford: Clarendon, pp. 242–244.
Davidson, Donald (1987). "Knowing One's Own Mind". *Proceedings and Addresses of the American Philosophical Association* 60, pp. 441–458. Reprinted in Davidson 2001.
Davidson, Donald (1990a). "Turing's Test". In *Modeling the Mind*, ed. by K. Mohyeldin Said, W. Newton-Smith, R. Viale and K. Wilkes. Oxford: Oxford University Press, pp. 1–11. Reprinted in Davidson 2004.
Davidson, Donald (1990b). "Representation and Interpretation". In *Modeling the Mind*, K. ed. by Mohyeldin Said, W. Newton-Smith, R. Viale and K. Wilkes. Oxford: Oxford University Press, pp. 12–26. Reprinted in Davidson 2004.
Davidson, Donald (1993). "Thinking Causes". In *Mental Causation*, ed. by J. Heil and A. Mele. Oxford: Clarendon Press, pp. 3–17.
Davidson, Donald (1994). "Donald Davidson". In *A Companion to the Philosophy of Mind*, ed. by S. Guttenplan. Oxford: Blackwell, pp. 231–236
Davidson, Donald (1995). "Could There Be a Science of Rationality?" *International Journal of Philosophical Studies* 3, pp. 1–16. Reprinted in Davidson 2004.
Davidson, Donald (2001). *Subjective, Intersubjective, Objective*. Oxford: Clarendon Press.
Davidson, Donald (2004). *Problems of Rationality*. Oxford: Clarendon Press.
Heal, Jane (1997). "Radical Interpretation". In *A Companion to Philosophy of Language*, B. Hale and C. Wright. Oxford: Blackwell, pp. 175–196.
Kim, Jaegwon (1984). "Concepts of Supervenience". *Philosophy and Phenomenological Research* 45, pp. 153–176.
Kim, Jaegwon (1998). "The Mind-Body Problem After Fifty Years". In *Current Issues in Philosophy of Mind*, ed. by A. O'Hear. Cambridge: Cambridge University Press, pp. 3–21.

Kim, Jaegwon (2003). "Davidson's Philosophy of Mind/Psychology". In *Contemporary Philosophy in Focus: Donald Davidson*, ed. by K. Ludwig. Cambridge University Press, pp. 113-136.

Lewis, Harry, A. (1985). "Is the Mental Supervenient on the Physical?" In *Essays on Davidson: Actions and Events*, ed. by B. Vermazen and M. Hintikka. Oxford: Clarendon Press, pp. 159-172.

Manor, Ruth (2001). "On the Overlap between Semantics and Pragmatics". *Synthese* 128, pp. 63-73.

Moore, George, E. (1922). *Philosophical Studies*. London: K. Paul, Trench, Trubner & Co.

Putnam, Hilary (1975). "The Meaning of 'Meaning'". In *Language, Mind and Knowledge*, ed. by K. Gunderson. Minnesota Studies in the Philosophy of Science, VII. Minneapolis: University of Minnesota Press, pp. 131-193.

Shagrir, Oron (2002). "Global Supervenience, Coincident Entities and Anti-Individualism". *Philosophical Studies* 109, pp. 171-195.

Shagrir, Oron (2009). "Anomalism and Supervenience: A Critical Survey". *Canadian Journal of Philosophy* 39, pp. 237-272.

Shea, Nicholas (2003). "Does Externalism Entail the Anomalism of the Mental?" *Philosophical Quarterly* 53, pp. 201-213.

Sider, Theodore, R. (1999). "Global Supervenience and Identity across Times and Worlds". *Philosophy and Phenomenological Research* 59, pp. 913-937.

Stalnaker, Robert (1996). "Varieties of Supervenience". *Philosophical Perspectives* 10, pp. 221-241.

4

Scientific Theory and Natural Language, Holism, and Measurement

ELI DRESNER

1 Holism and the Continuity between Natural Language and Scientific Theory

Meaning holism is a thesis that gives rise to disagreement even with respect to its exact statement, let alone its truth (Pagin 2006). Here it will be construed as the view that the meaning of linguistic expressions is constituted by a network of relations that they have among them. (See Dresner 2002 for elaboration.) Different holistic views of meaning characterize these meaning constituting relations in distinct ways: According to some positions they are truth-functional, according to others inferential (where inference is defined without appeal to truth), and in yet other theories, which ground linguistic meaning in propositional thought, they are construed as functional relations that obtain among the constituents of a given thinker's mental economy (Block 1986).

The thesis of meaning holism is thus a constituent of several distinct conceptions of linguistic meaning, and therefore the arguments in support of the thesis are variegated. Some arguments present holism as a consequence of the defused character of inter-personal interaction that gives rise to meaning: Expressions are assigned meaning in a process

Hues of Philosophy.
Anat Biletzki (ed.).
Copyright © 2010.

(e.g. interpretation) in which many other expressions are assigned meaning interdependently (Davidson 1984). Other arguments, already hinted at above, see meaning holism as arising from the functional nature of our internal mental states (Block 1986). As attested by Fodor and Lepore's (1992) review of holistic positions and the many reactions it has given rise to, the mapping and assessment of the various arguments for meaning holism is a formidable task. It will not be taken up here.

Similarly, the objections against meaning holism are the subject matter of heated debate. Three such key objections consist in claims that holism is inconsistent with characteristics that natural languages clearly manifest, namely their being compositional, learnable and stable. The first objection is to the effect that intersentential meaning imputing connections have to exclude the determination of the meaning of a sentence by its composition from sentence parts. The second, having to do with learnability, sees unresolvable tension between the holistic picture of language and the incremental process through which we break into our first language that seemingly requires our acquisition of independent meanings one after the other. The third objection depicts holistic semantic interdependence among linguistic expressions as implying that every local change in the network that constitutes meaning must result in a change in the meaning of all expressions in the language. These objections will not be discussed here; the first has been argued against, e.g. in Pagin (1997) and the two others in Dresner (2002 and 2003).

As is the case in many philosophical disputes, then, there is no widespread agreement on the value of the arguments in favor of meaning holism, nor on the strength of the objections against it. This state of affairs leaves the door open for further, indirect considerations that may bear upon the question and possibly support either the holistic camp or the one that opposes it. One kind of such considerations, which purports to support a holistic view of linguistic meaning, appeals to a comparison between natural language and scientific theory.

The most direct way of making use of an alleged analogy between natural language and scientific theory is through an argument such as the following. First, scientific theories are holistic (in the relevant, semantic sense). Second, scientific theories and natural language are essentially the same—they are merely distinct areas within a single linguistic network and therefore manifest the same basic characteristics (while exhibiting various important differences that should be acknowledged). Therefore, meaning holism obtains for natural language as well. This may not be as evident as in the scientific case, but holds just the same.

Has such an argument ever been explicitly presented in support of meaning holism? Probably not. As noted above, arguments for holism are typically embedded within specific conceptions of meaning, while the above so-called "argument from the continuity with science" is not. However, Quine seems close in spirit to this argument. Quine (1951) famously argues for verification holism with respect to scientific theories (The Quine-Duhem Thesis), which, if conjoined with a verificationist conception of meaning, would entail a holistic view of meaning with respect to scientific theories—i.e. the first premise in the above argument. Quine certainly accepts that second premise of the argument, namely that there is continuity between scientific theory and natural language (see the discussion below), and therefore meaning holism of natural language follows. However, Quine cannot be claimed to endorse a verificationist view of *meaning*—because he disavows meanings altogether—and therefore the suggested argument cannot be ascribed to him. Here is the point in his own words (Quine 1986: 155-156): "I find it [verificationism—e.d.] attractive. The statement of verificationism relevant to this purpose is that 'evidence for the truth of a sentence is identical with the meaning of the sentence'; and I submit that if sentences in general had meanings, their meanings would be just that. It is only holism itself that tells us that in general they do not have them."

It is thus difficult to find an explicit affirmation of the argument in question—in Quine, or elsewhere. As for a rejection of the argument, things are different: Dummett considers it explicitly, and then dismisses it as erroneous. Not surprisingly, it is not with the first premise of the argument that Dummett finds fault: He wholeheartedly endorses verificationism about meaning. He is willing to accept verification-holism with respect to scientific theories, and therefore grants meaning holism in the scientific context. However, Dummett rejects the second premise of the argument, namely that there is continuity between scientific theory and natural language. Here is what he says (Dummett 1991: 232):

> We may distinguish between a narrow and a broad sense of the term 'theory'. A theory in the broad sense is any deduc-

tively organized body of explanatory hypotheses (even this definition excludes group theory and the like). A theory in the narrow sense is one containing at least one theoretical term. We may take a theoretical term to be one whose significance, as it occurs in statements of the theory, depends solely on its role within the theory, and not at all on its use, if any, in extra-theoretical statements, and one which, furthermore, does not, according to that theory, stand for anything that is even in principle directly observable. As a thesis about scientific theories, in the narrow sense, the Duhem-Quine thesis is credible enough. There is not enough continuity, however, between the sector of language used in a scientific theory and other sectors of language to justify extension of the Duhemian principle to the use of language as a whole.

The expressions of natural language, Dummett argues, are not theoretical: They are not internal to any theory, and they do not stand for unobservables (as theoretical terms do). Should these two claims be accepted? The first seems problematic: If the whole of natural language is conceived of as a single unified theory (our intuitive, everyday theory of the world, so to speak), then surely all the terms that appear in it are internal to it—there is nowhere else for them to appear. And what about the second characterization of theoretical terms (that they stand for unobservables), and the claim that it does not apply to natural language expressions? It too can be contested, and will be returned to later in this section and in the following ones.

Dummett (1991: 232-234) acknowledges that there is diffusion and permeation back and forth between natural language and science: Scientific terms often have their roots in everyday expressions, on the one hand, and enter everyday language together with some of their theoretical baggage, on the other hand. However, Dummett argues, this does not undermine his position. As for the first type of connection of the two just mentioned (science being grounded in natural discourse), it does not entail that natural discourse shares with science its theoretical character (234): "The sense in which the [everyday—e.d.] concept of distance is embodied in a theory is that to have the concept is to make a number of general assumptions and to engage in a network of interrelated practices; but these assumptions lack the hypothetical character that may be attributed to even the best established scientific theory." Connections of the second type (where scientific discourse appears within natural language) are superficial, says Dummett, and mask the self-standing nature of science vis-à-vis everyday language: Changes in scientific theory will not affect everyday discourse that includes expressions and statements extracted from such theory. Are these two assessments of Dummett's correct? This question too will be returned to below.

It may be asked whether Dummett's objection to a (partly) hypothetical argument for meaning holism is the only context where the connection between scientific theory and natural language comes to bear on the debate over meaning holism and on the philosophy of language in general. If the answer to this question is positive, then the interest in studying this connection would be limited. However, I argue that the said connection, while admittedly not playing an explicit role in arguments for holism, nevertheless serves important functions in philosophical accounts of meaning that are holistic in nature, or that manifest characteristics which are closely related to holism. Here are two key examples.

1 *Interpretation and Translation as Theorizing.* Davidson famously construes radical interpretation as the construction of a *truth theory* for the utterances of a given speaker. Borrowing Dummett's terminology (as it appears in the passage quoted above), I argue that the term "theory" is used here in the narrow and stronger sense, and that therefore Davidson implicitly analogizes interpretation to scientific theory construction. Here is why. It is not the case that according to Davidson interpretation merely consists in the production of a "deductively organized body of explanatory hypotheses" (this is Dummett's definition of the wide and weak sense of "theory"). This would be consistent with a bottom-up conception of interpretation (that Davidson eschews), according to which syntactically primitive expressions are interpreted on their own. Rather, an interpretational truth theory consists in the assignment of truth conditions to sentences and references to some sentence-parts on the basis of a class of observations which the theory faces as a whole. The analogy with scientific theory is thus substantial and clear.

According to Davidson, then, the ascription of meaning is on a par with scientific theorizing. Such ascription, in turn, constitutes meaning.

Therefore it follows that not only is meaning-ascription a theoretical endeavor. Rather, the result of such ascription, i.e. linguistic meaning, is theoretical in nature as well—that is, it consists in a hookup between language and the world that is like the hookup that obtains between scientific theory and the world. The connections that obtain between words and things, as well as truth conditions assigned to sentences, are in both cases—that of everyday language and that of scientific theory—dependent on a match between a theory and body of evidence. Thus, although Davidson never relies on the analogy between scientific theory and natural language and is famously opposed to there being a continuity between hard science and everyday psychological talk, the analogy just stated between scientific theory and meaning ascription is, nevertheless, an underlying principle of his outlook.

Going back to Quine, we can see that the situation is the same. Quine conceives of radical translation as a theoretical endeavor, and expresses this conception in his account of translation through the use of such terms as *analytical hypotheses*. The result of translational theorizing is a hookup—this time between languages—that is the same as that which obtains between a scientific theory and the aspects of the world it aims at accounting for. So here too what is viewed in linguistic meaning as in need for an account—i.e. translation—is construed in analogy with scientific theory.

2 *First Language Learning and Scientific Theorizing*. Quine's interpersonal, translational perspective on linguistic meaning is coupled with a first person perspective that is similarly modeled after scientific theorizing. According to Quine (1974) the only thing we have to go on in the dual process of learning our first language and forming our view of the world around us is sensory stimulation. On the basis of regularities in sensory stimulus we stipulate the existence of everyday objects, together with their properties and the relations that obtain among them. Thus our everyday ontology is on identical footing with that which scientific theorizing gives rise to. The child breaking into her first language and forming a conceptualization of the world around her is therefore conceived by Quine as engaged (albeit unconsciously) in a project that is essentially identical to that of the scientist: The child is forming a theory (and is thereby stipulating an ontology) that best fits the evidence. Thus this major

feature of Quine's philosophy, too, depends on there being continuity between everyday language and science.

One of the more concrete issues upon which this view of language learning bears upon, which has already been touched upon above, is that of conceptual change through early childhood. As may be recalled, one of the consequences of holism is the possibility of changes in meaning that arise from changes in the network of connections that constitute meaning—this is viewed by some critics of holism as implying improbable instability of linguistic meaning. Thus if the child's engagement with language and world is viewed as theoretical and therefore holistic in nature, then we may expect changes in his concepts that arise from changes in the theories they are embedded in—theories that develop and evolve together with the child's changing capabilities and interests. According to opposing conceptions of early development of conceptualization and language, on the other hand, more stability may be expected: Basic hookups (e.g. between words and objects) are formed and then stay fixed. Which of these conceptions is more in accord with empirical evidence concerning early conceptual and linguistic development? As noted in Dresner (2002), different researchers give opposing answers to this question, but there are certainly those who argue that the Quinian framework should be preferred over its counterpart.

It is certainly not my purpose here to enter into the debate over the nature of first-language learning and early conceptual development. Rather, I present this debate in support of the main claim of this section, namely that the question whether natural language is continuous with scientific theory is important for the study of language in general, and for the debate over meaning holism in particular. This example and the previous one show that the view that such continuity exists, even if only seldom playing an explicit role in philosophical arguments, nevertheless underlies (and is implied by) major philosophical perspectives on language, as well as approaches to its empirical study. Therefore an assessment whether such continuity does indeed obtain (or, as Dummett argues, does not) is called for.

2 Measurement as an Intermediary between Natural Language and Science

How should an assessment of the continuity (or lack thereof) between natural language and scientific theory be undertaken? The discussion in the previous section indicates that an independent examination of each type of discourse, followed by a comparison between them, is not a viable option: We saw that the comparison between natural language and scientific theory is supposed by many philosophers to enhance our understanding of the former, and therefore such understanding cannot be presupposed in an assessment of the affinity between the two.

An alternative course, which will be taken up here, is to consider contexts where natural language and scientific theorizing interact with each other. It is there that we should expect to find signs of discontinuity and friction if indeed scientific theory and natural language function in fundamentally different ways, or, on the other hand, where we may find indications of basic underlying similarity. Notions that seem to appear both in science and in everyday language offer us an opportunity to observe whether they take part in two distinct games, or a single continuous one.

Probably the most natural choice for this purpose would be to consider the basic measurable properties, such as length, mass or temperature. These play a key role in our everyday life and everyday language, and have been doing so for centuries (temperature) and millennia (length). On the other hand, they serve as the cornerstone of modern science, by enabling the formulation of mathematically couched laws that involve ever more abstract quantifiable properties and increasingly complex connections among them. This whole edifice is erected on the basis of measurements that are employed in contexts that we would not call scientific.

Indeed, Dummett seems to acknowledge (albeit implicitly) that a consideration of the way we talk and think about the basic measurable quantities is suitable for examining his thesis that there is discontinuity between scientific theory and natural language. As may be recalled from the previous section, he considers explicitly the notion of distance and claims that the way it is put to use in everyday talk shows that it is indeed non-theoretical, as opposed to theoretical terms employed in science.

Is Dummett justified in his claims concerning distance? I argue that before this question can be answered an important methodological point has to be made. Dummett appeals to our intuitions concerning distance, but ignores a whole body of knowledge that is based upon these intuitions, refines them and develops them. This body of knowledge is the Theory of Measurement: The systematic study of (i) the conditions that allow numeric measurement in some contexts and not in others, (ii) the various types of measurement, and (iii) various philosophical and methodological problems that arise with respect to measurement. Although measurement has been practiced for thousands of years, this domain of inquiry has come into existence only towards the end of the nineteenth century, through the work of Helmholtz (1977) and Holder (1901). It has been pursued (albeit never widely) during the first half of the twentieth century, e.g. by Campbell (1928) and Stevens (1946), and has reached maturity in the form of contemporary Representational Theory of Measurement (RTM), the foundations of which have been laid by Suppes and his collaborators (Krantz, Luce, Suppes and Tversky 1971–1989).

In the remainder of this section I propose to present a brief outline of the representational theory of measurement. Then, in the next section, I use it (and the philosophical debates that surround it) in order to address the question this paper is concerned with, namely the connection between natural language and scientific theory. Before doing so, let me note in passing the affinity between the theory of measurement and logic. Both domains of research are concerned with the study of preexisting practices (measurement and inference, respectively) and both employ formal means. Moreover, both domains may be labeled methodological, having both philosophical and nonphilosophical aspects. Thus, I see the appeal to measurement theory as one of the ways in which this paper is related to the work of Ruth Manor, to whose memory it is dedicated.

The main subject matter of the representational theory of measurement is *fundamental* measurement—any association of numbers with objects in order to designate and describe their properties that does *not* presuppose a prior such association. Length and mass measurement are central examples of fundamental measurement, while the measurement of velocity, for example, depends on prior measurement of distance and time, and is therefore *derived* rather than fundamental measurement.

According to RTM, fundamental measurement consists in a homomorphism—a structure preserving mapping—between a domain of physical objects and the numbers (typically the real numbers). The former, physical domain is called in measurement theory an *empirical relational structure* (more on this term below, in the following section). It consists of physical objects and several primitive qualitative relations among them (and operations on them) that are specific to the quantity that the empirical structure in question allows measuring. For example, the empirical relational structure that underlies length measurement includes the basic relation of qualitative length comparison (that can be operationally realized through the juxtaposition of any pair of objects one next to the other, and observation which of them extends the other), and the operation of length-oriented concatenation (i.e. putting one object right after the other).

An empirical structure of this kind may be observed (or stipulated—see below) to satisfy various formal axioms. Thus the length comparison relation can be readily acknowledged to be an ordering relation and the concatenation operation can be seen to be, for example, commutative. If the said structure satisfies several certain axioms of this kind it can be mathematically proved that there is indeed a structure preserving mapping from it to the numbers. That is, there can be proved to exist a function that assigns a number to each object in the domain, and also assigns a numerical relation (operation) to each basic empirical relation (operation) in the domain, such that the following condition is satisfied: Any pair (or n-tuple) of physical objects in the empirical domain satisfies an empirical relation if and only if the numerical measures assigned to these objects satisfy the corresponding numerical relation. The proof that such a function exists is called in measurement theory a *representation theorem*.

Note that in virtue of this mapping (in particular, in virtue of the "only if" part in its definition) measurement serves us well for what Swayer (1991) calls "surrogative reasoning". We assign numbers to objects, say in length measurement, and then we reason arithmetically about the numbers assigned. The results of such arithmetical reasoning can be applied backwards to the physical domain just because the representation theorem assures us that, indeed, the physical domain is in complete correlation with the numerical one. Thus, according to this picture of measurement, the realm of numbers and the physical world are kept apart, yet the structural correlation between them renders the former so highly useful in our dealing with the latter.

Also, note that in order for measurement to be established the empirical domain need not be isomorphic to the numerical one; it need only be homomorphically embedded in it. Therefore measurement can make variegated representational use of the rich structure of the numbers. This gives rise to a variety of measurement scales, among which can be found, for example, the ratio scale (that characterizes length and mass measurement) and the interval scale (used, e.g. in temperature measurement). In the former case (ratio scale) the properties of the number 0 are representationally meaningful, while in the latter case they are not. This is expressed by the fact that in the latter case (e.g. in temperature measurement) the decision which object(s) should be assigned 0 is conventional rather than substantial. (Let us ignore the discovery of the absolute zero in this discussion.) Thus different measurement types (scales) allow different classes of legitimate transformations on the representing numerical domain—i.e. functions that match numbers with other numbers in a way that retains the representation of the given empirical structure. A theorem that characterizes the class of such transformations in each given context of measurement (transformations that are called, somewhat confusingly, changes of scale—e.g. from Fahrenheit to Celsius) is called in measurement theory a *uniqueness theorem.*

This very brief outline of the representational theory of measurement glosses over many details and surely leaves many questions unanswered. Some of these details and questions will be discussed in the next section.

3 Measurement Theory and Natural Language

I turn now to three different issues having to do with the practice and theory of measurement, as just outlined. The discussion of each of them implies that, in fact, a close examination of measurement supports the thesis that natural language and scientific theory consist in a single and continuous linguistic-theoretic fabric.

1 *Inventing Temperature*

Where is the cut off (if indeed there is one) between natural discourse and scientific theory in the domain of temperature? Some would probably say that theoretical thought is introduced into this domain only when the phenomenon of temperature is given theoretical explanation. Thus, for example, the theory of phlogiston was developed in order to account for temperature, and now, after this theory has been rejected, the theory of molecular mechanics serves the same function. And what about the numeric measurement of temperature itself? Is it a part of science, or of everyday discourse? On the one hand, due to the quantitative nature of temperature measurement, as well as the fact that it is only a few centuries old, one may want to include it in science. On the other hand, the fact that temperature measurement preceded theorizing of the kind just mentioned, as well as the relative constancy of such measurement in the face of theoretical change, seem together to imply that temperature measurement itself is proto-science, a part (albeit an exact part) of our everyday scheme—couched in everyday language—of thinking and speaking of the world.

In his fascinating book, *Inventing Temperature*, Chang (2004) provides ample proof that the first of these options is closer to the truth and that measurable temperature itself is theory laden. (Indeed, the title of the book is designed to suggest just that.) Or, rather, what Chang shows is that the very opposition between the two options is misguided because no cut-off between science and natural language can be found if the history of temperature measurement is carefully studied and the philosophical issues it raises are considered. A key discussion that demonstrates this proposition is that of *fixed points* in temperature measurement (Chang 2004: 39-44).

As noted in the previous section, temperature is measured on an interval scale. This means that the choice which temperature is to be assigned 0 is conventional, as is the choice of unit of measurement. Thus the fixation of two points is needed: One will be the starting point of the scale, and the other will mark the end of an interval that is to be divided into an agreed-upon number of temperature units. But how can fixed points be found? In order to verify that a certain object remains in a fixed temperature (or that events of a given physical type always happen at the same temperature) we need temperature measurement, but such measurement is the very thing which fixed points are needed for. As noted by Chang (2004: 40), Newton suggested that the temperature of human blood (presumably while within the living body) is fixed; how could he have known that the human body changes its temperature before measurement was possible?

The solution to this problem was to use (what was only later called) a *thermoscope*. A thermoscope is a proto-thermometer: It contains a liquid of some kind that may be observed to contract or expand with the change of temperature. Such a device yields *ordinal* measurement: It can tell us if one body is warmer or colder than another (or whether the same body has become warmer or colder), but these pronouncements have no quantifiable meaning. (According to contemporary representational theory, what is appealed to in this kind of measurement is only the order of the numbers.) Through the use of a thermoscope fixed points can be found, and on their basis measurement may proceed.

It could be further inquired how the use of a thermoscope can be justified. How do we know that liquids expand when heated, and contract when cooled? Chang's answer is that we have nothing to rely on but our ability to correlate changes in liquids' volumes with our tactile perception of heat and coldness. Perceptions of this very basic and rough kind are the starting point of the whole chain of reasoning that leads to fullfledged numerical measurement of temperature. This is not to say, though, that throughout this process heat perception retains the status of unerring observation, with which all later temperature measurement must conform. Rather, we are sometimes willing to disregard perception when it contradicts judgments we want to hold on to. For example, we uphold the view that contiguous liquids have the same temperature throughout (if given enough time for temperature to equalize), and this in the face of well-known contradictory sensory stimulation. Also, we often give priority to numerical measures of temperature over our feelings. There is equilibrium between the basic sensory pronouncements and more sophisticated judgments that cannot be described as simple logical dependence.

Chang's objective in developing the analysis just outlined is historical, as well as epistemological. I propose to use it here for semantic purposes. Having reviewed this analysis we may now return to the question: When is theoretical thought introduced into temperature measurement? The answer, I think, is clear: from the start. We saw this demonstrated in two ways. First, the very idea of temperature as an enduring property of objects was shown above to be a theoretical construct: Like good scientists we uphold it against contradictory experience by making changes in our theory of temperature perception. And second, our very access to fixed temperature, which is necessary for temperature measurement, depends on a theory that relates volume to temperature. In order to measure temperature we do not directly observe if a body changes its temperature or not—we need a theory to support judgments of this very basic kind.

Thus, I argue, the consideration of temperature imputes content and gives credence to the claim that everyday concepts are theory laden. Is temperature theory laden in the weak sense of the two invoked by Dummett, i.e. in the sense that to have it is merely "to make a number of general assumptions and to engage in a network of interrelated practices"? I think not, at least if the expression "general assumptions" is to be construed so as to oppose "theoretical hypotheses", as Dummett obviously intends it to be. As we saw, the practice of temperature measurement presupposes a theory—i.e. judgments that are verified not by straightforward bottom-up processes, as Dummett envisions the verification of natural language sentences, but through relations to evidence that are on a par with those that scientific theories have. This is true for theories that attempt to explain what heat is, for the practical tasks of finding fixed points, and for the very idea of temperature as a persistent property.

It should be noted, again, that the claim being made here with respect to temperature, and that will be argued for below through the consideration of other measurable properties, is not presented as capturing a characteristic that is unique to conceptualization and language in the domain of measurement. Rather, our question was whether there is a fault line between scientific theory and natural language in general, and the discussion of temperature was presented here as providing support to the view that there is no such fault line. We saw that there is no point where temperature becomes theoretical: It is so from the start, and arguably becomes more elaborately and explicitly so when our understanding of it improves and when it starts to play a role in quantitative science. Thus the role played by temperature in quantitative science motivates valuable work such as Chang's, which unearthes the theoretical nature of the concept in proto-science and in everyday discourse. However, this is not to imply that only concepts that become enmeshed into quantitative science are theoretical in this way. Rather, most plausibly the theoretical baggage of other concepts simply does not receive attention and scrutiny as that of measurable properties such as temperature. Therefore, I conclude, the consideration of temperature measurement does indeed serve to expose an overall unity between science and natural language.

2 *Inventing Length*

One could argue against the above considerations in the following way. Temperature is a distinctively theory-laden concept, and therefore it is no surprise that we cannot detach its natural, prescientific part from its theoretical content. But things are not the same with other concepts, such as length. It is thus no coincidence that Dummett appeals to length rather than temperature as exemplifying his claim that scientific theory and natural language are different. However, as I turn now to show, a closer examination of the underpinnings of length measurement indicates that here, too, theoretical and natural discourse cannot be separated.

As may be recalled from the previous section, measurement requires (or implicitly presupposes) a proof that there exists a homomorphism from a given empirical structure to the real numbers. In order for such a proof to be carried out, the physical, empirical structure needs to be construed as satisfying a certain set of axioms. In the case of length measurement the empirical structure involves the primitive relation of qualitative length comparison between objects (realized through their juxtaposition one next to the other), and the operation of length concatenation on objects (realized through their juxtaposition one after the other). The axioms that this structure needs to satisfy in order for a representation theorem to go through include the following:

(i) Length comparison should be a weak ordering (in the mathematical sense).
(ii) The domain of objects should be closed under the operation of concatenation (i.e., the concatenation of every pair of objects should exist, actually or hypothetically).
(iii) Length comparison should be Archimedian with respect to length concatenation. (That is, if you take any two objects, the concatenation of the shorter one to itself a certain finite number of times would be longer than the other object of the two).

A complete set of the required axioms can be found, e.g., in Bozin (1998).

But how do we know that the empirical structure that underlies length measurement indeed satisfies these axioms? How can we justify the claim that it does? This is a key question, and the term "empirical relational structure" suggests an answer to it that is presupposed by many proponents of the representational theory of measurement. The implicitly suggested answer is that our direct observations of objects (and of the above described relations and operations on them) can provide sufficient grounds for the claim that the physical structure satisfies the formal axioms. Indeed, Carnap himself holds that this is the case. In his later (1966) view of the way observation gives rise to science, Carnap gives extensive attention to measurement as a junction in which quantitative language and thought are introduced. His explication of measurement (e.g. of length) invokes analogues of the so-called empirical structures that we have been discussing here, and he argues that their properties can be completely grounded in observation and in our measurement practices (Carnap 1996: 100).

However, is this view justifiable? Is our experience with objects sufficient to ground the above stated axioms? In a recent article, Batizky (1998) argues forcefully that this is not the case. His argument, in a nutshell, is this. First, the measurement of, for example, length, requires the empirical relational structure to satisfy axioms that are stronger than those entertained by Carnap. (This is what modern measurement theory teaches us.) Second, these stronger axioms cannot be directly supported by our direct observations. The move from actual experience to the far-reaching content of the axioms goes beyond the innocent generalizations (if indeed there are generalizations that are innocent in the required sense) that Dummett relegates to nontheoretical discourse. For example, consider again the Archimedian axiom. How can our everyday experience justify the claim (which this axiom boils down to) that there are no infinitely large or infinitely small objects? Surely our length-related practices are geared only towards objects that are indeed commensurable with each other, but why should there not be objects that are incommensurable in the vast domains of the universe that are beyond our practical reach? The claim that, indeed, there are no such objects cannot be viewed as a legitimate extrapolation of our experience that a staunch empiricist may be able to defend.

It would seem, then, that the empirical relational structures that are postulated by the representational theory of measurement are not empirical in the strict empiricist sense. Is there any other sense in which they can be said to be empirical? Possibly only in that they are structures involving physical objects and relations that are available to us through experience, as opposed to the purely mathematical structures that are used to represent them. But is not this weak reading of "empirical" detrimental to the theory of measurement? Does not the theory require for its coherence and fruitfulness that the so-called empirical relational structures it stipulates be grounded bottom-up in experience, and only then be homomorphically mapped into the abstract and experience-transcendent domain of the numbers?

I argue that the answer to these questions is negative. Contrary to the (often implicitly held) view of many proponents of the representational theory of measurement, the represented structures that the theory stipulates *cannot* be the product of direct observation (as has been argued above), nor *need* they be such products for the theory to be fruitful and coherent. Instead, these structures should be thought of as *theories* that underlie our thoughts about (and dealings with) objects and their lengths. In the theory of measurement these basic theories are given clear and explicit expression, possibly for the first time, in the form of the axioms that govern the so-called empirical relational structures. However, these theories are not created by measurement theorists in the same way that higher-level physical theories are introduced by physicists. Rather, these theories—such as the one that underlies length measurement—are implicit in our everyday talk, e.g. of length and distance, and in our everyday practices.

The conclusion of this discussion for the foundations of measurement is as follows. We need not and should not think of the general formal structure offered by RTM as involving a structure preserving mapping from a concrete, theory-free domain into an abstract, numerical one. Instead, measurement theory can be placed inside Nuerath's boat, or within Quine's holistically-interconnected sphere of theory and language. Viewed this way, measurement consists in a mapping from a lower-level theoretical construct to the numbers. Measurement is not the hookup between nontheoretical discourse and theoretical physics, but rather a step in the elaboration of an approach to the world that is theoretical from the start. This outlook does not underplay the importance of measurement, nor does it detract from the significance of measurement theory in elucidating the theoretical underpinnings of this practice.

The result of the discussion as regards the main question we are concerned with here is this. Contrary to first appearances, the concept of length (or distance) is theory-laden as well. Indeed, measurement theory helps uncover the lower-level theory that underlies the practice of length measurement. Thus the consideration of this concept, too, supports the claim that there is no clear-cut division between natural language and scientific theory with respect to theoretical involvement.

3 *The Identity of Measurable Properties*

What makes a certain measurement procedure a measurement of length rather than of mass? How can two distinct procedures measure the same property? And how is the theory of measurement related to the identity conditions of measurable properties? Providing full-fledged answers to these questions is beyond the scope of this paper, but I propose to consider several such answers to them and show that the more plausible ones must rely on there being continuity between natural and scientific language.

One kind of answer to the above set of questions may be called Platonist. According to views of this kind there are language- (and thought-) transcendent metaphysical facts of the matter what objects and what properties there are, and, in particular, what the measurable properties are. Viewed this way, distinct measurement practices can easily be construed as measuring the same properties, and the theory of measurement tells us what distinguishes classes of measurable properties (such as the class of lengths) from other, less structured classes. Proponents of this position include Swoyer (1991) and Mundy (1987), as well as philosophers like Armstrong (1989), who subscribe to the general outlook that metaphysics can play the role of a first philosophy.

I will not attempt here to argue against this position. Suffice it to say that it is unappealing to those philosophers who view language and thought (in whatever explanatory order) as not merely mirroring a world that is "carved at its joints" independently of our thinking and talking about it. Such philosophers, to whose views I subscribe, see properties as inextricably related to concepts and predicates, and therefore maintain that there are stronger, constitutive connections between properties and the theories and practices that we apply in assigning these properties to objects and in reasoning about them.

If this second, thought-centered (or language-centered) perspective is adopted, there is still an important choice that is left to be made (either explicitly or implicitly) regarding the scope of uses that are to determine the meaning of measurement predicates, and thereby the identity of the measurable properties. (The term "uses" is used here in the widest and most permissive way, including both practical uses and so-called cognitive uses in theorizing and reasoning.) One restrictive option is to view measurement theory as exhaustively articulating the contexts that impute meaning to the measurable predicates. This is in tune with the way higher-level scientific theories (such as particle physics) seem to completely determine the meaning of the terms that are introduced in them. The second option is to allow that many mundane contexts of discourse that are not captured by measurement theory are essential to the meaning of the basic measurable predicates (such as length and mass) and to the identity of measurable properties. In what follows I briefly explore some of the difficulties that arise if the first, restrictive route is taken, and conclude that due to these difficulties the second should be preferred.

The restrictive approach seems, at first, to be appealing. It construes the primitive relations and operations that constitute the empirical relational structure for a given measurable predicate as exhausting its meaning. According to this perspective the notion of length, for example, is derived from qualitative length comparison and/or length concatenation as primitives that are related to measurement practices on the one hand and satisfy axiomatic formal restrictions on the other hand. (Note that an extreme version of this view consists in an outright operationalization of the measurable properties, tying them to the concrete procedures associated with the primitives of their underlying structures. This version suffers from obvious difficulties, such as its ruling out the possibility of the same property being measured in distinct ways. A more moderate version of the view takes a measurable property as essentially tied not to, for example, a given procedure of comparison, but rather to the ordering that results from this procedure (but could have been established in an alternative way).)

However, as is made evident by attempts to elaborate this line of thought, it is inherently problematic. As an example, consider Ellis's (1968: 75-89) claim that it is only the length-theoretic *ordering* relation among objects (which qualitative length comparison helps establish) that is essential to the notion of length. It is a consequence of this claim that concatenation operations that are distinct from the standard one may be used in length measurement, as long as they satisfy the required axioms. Thus, says Ellis, perpendicular concatenation is consistent with the concept of length no less than standard concatenation. (Such alternative concatenation yields a different assignment of numbers to objects as their lengths, but this assignment, too, has no more and no less merit to it than standard assignments, e.g. the metric one.) So what is the justification for our using the form of concatenation that we do? Ellis stresses his opinion that the familiarity of standard concatenation and the ease with which we use it do not provide such justification and argues, instead, that the only solid grounds for preferring one concatenation scheme over the other have to do with simplicity and convenience in physical theorizing. Ellis, we see, is an unproclaimed ally of Dummett's in one respect: He divorces scientific theory from everyday discourse. However, on another important count, he can be said to oppose Dummett: Ellis places the concepts of measurement in the domain of science, while Dummett views them as standing on the other side of the fault line, i.e. as a part of natural language.

Is Ellis's position plausible? I argue it is not. Ellis is not satisfied with merely suggesting that his alternative notion of concatenation gives rise to an alternative notion of length, and that this alternative notion might play the role of the traditional notion in some theoretical contexts with equal success. Had he done so there might have been no room for disagreement, although the onus would have been on him to show how his alternative notion of length can be incorporated within the whole body of physical theory. However, Ellis argues that the connection of our *traditional* notion of length to standard concatenation is weak and inessential, a matter of mere familiarity and habit, or possibly (pre)historical accident. But this does not seem to be the case: Surely our choice of concatenation procedure is not conventional and arbitrary in the same way that, for example, our choice of the measurement *unit* is, as Ellis's analysis suggests. Rather, our choice of the traditional concatenation relation seems to be related, among other things, to the fact that we often move around the world in what seem to us as straight lines (because it is most efficient) and that we view distances along such straight lines as concatenations of measurement units. Similarly, it may be the case that when we deal with objects we place them more often along straight lines rather than at right angles, and that this, too, has to do with the way we concatenate lengths. Although I cannot present here a complete account of the way our practices affect the way we concatenate lengths, I argue that such an account can and needs to be made, and that therefore Ellis is wrong in holding that length is defined in terms of order alone.

We see, then, that Ellis's "lean" concept of length yields untenable results with respect to our everyday conception of length and thus should be rejected. Moreover, the same considerations also show that grounding the notion of length in length ordering *together with* length concatenation (i.e. the whole empirical structure that underlies length measurement) is not viable either. The reason is that in this case, also, we are left with too weak a connection between the would-be definition of length and our everyday dealing and reasoning about it. According to the suggested variant of Ellis's position, the definition of length matches our everyday dealings with it and talk about it only contingently, and could have deviated from them without inconsistency (as Ellis's alternative suggests). Can there be such a gap between our use of the term "length" and what length is? Dummett would certainly give a negative answer to this question, and so should we.

Another way of making the same point is this. If the empirical relational structure of length is viewed as an implicit exhaustive definition of the notion, then the pronouncements of this structure are ascribed the status of necessary truths about length and other length-related judgments, some of which are of a very basic nature, are demoted to the lower status of contingent claims. Such a distinction, however, is implausible, and this not only because it invokes something like the analytic-synthetic distinction that has been famously and forcefully (Quine 1951) argued against. (As Quine's rejection of the distinction is closely associated with his holism, relying on it here would result in vicious circularity.) Rather, as Dummett would agree to as well, the definition is faulty because it attempts to define the notion of length through the use of an axiomatic structure that connects this notion top-down with higher physical theory, without regard to the basic assumptions and practices that ground it bottom-up and thus impute it with meaning. For example, the notion of length seems to be conceptually related to such more basic concepts as *distant*, which, in turn, may have conceptual ties to the concepts of *movement* and *time*. (A distant object is one that it takes more time to reach.) As opposed to the self-standing treatment of each measurable property in measurement theory, in everyday thought and language they are arguably tied together. As another example, consider mass and temperature: These concepts are surely related to tactile sensations that we have (of heaviness and heat) that are not captured by the empirical structures that underlie these concepts according to measurement theory. As already said above, it is beyond the scope of this paper to articulate the ties of measurement concepts to everyday practices and judgments but, nevertheless, I argue that such ties do contribute to the meaning of these concepts.

Thus the claim that the empirical relational structures stipulated by measurement theory consist in exhaustive implicit definitions of the properties they underlie should be rejected, and this—by philosophers of language of both holist and nonholist persuasions. In view of the above considerations, measurement theory is better thought of as capturing the structural aspects of several preexisting and rich concepts, aspects that allow these concepts to be involved in numeric measurement. Furthermore, the theory indicates what is required from newly-formulated concepts in order for them to be measurable as well. The fact that measurement theory does not give us a full account of what length or mass are does not detract from its utility and importance.

Nothing in the foregoing discussion applies specifically to length, and therefore its conclusions apply to all basic measurable predicates (as opposed to those that are theoretical constructs). Thus the attempt to draw a clearly-marked line between natural language and science in a way that includes the measurable properties within theoretical science and excludes them from everyday discourse is unsuccessful. Such a cut disassociates the measure concepts from their meaning base in a way that amounts to treating them as artificially coined terms of art (like electrons and quarks), which they are not. In the previous two parts of this section, on the other hand, we saw that trying to draw the line the other way, so as to include the measure predicates in natural language and divorce them from theoretical science, fails as well. Together these two results tell us that there is no line to be drawn. It can be concluded that a close consideration of measurement theory supports the view that scientific theory and natural language form a single continuous whole. And, to the extent to which scientific theory is acknowledged to be holistic, these considerations also support a holistic view of linguistic meaning.

References

Armstrong, D. (1989). *Universals: an Opinionated Introduction.* Boulder, Colorado: Westview Press.
Batitsky, V. (2000). "Measurement in Carnap's late Philosophy of Science". *Dialectica* 54, pp. 87–108.
Block, N. (1986). "Advertisement for a Semantics for Psychology". *Midwest Studies in Philosophy* X, pp. 615–678.
Bozin, D. (1998). "Alternative Combining Operations in Extensive Measurement".*Philosophy of Science* 65, pp. 136-150.
Campbell, N. (1928). *An Account of the Principles of Measurement and Calculation.* London: Longmans.
Carnap, R. (1966). *Philosophical Foundations of Physics.* New York: Basic Books.
Chang, H. (2004). *Inventing Temperature.* Oxford: Oxford University Press.
Davidson, D. (1984). "Radical interpretation". In Donald Davidson, *Inquiries into truth and interpretation.* Oxford: Clarendon Press, pp. 125–139.
Dresner, E. (2002). "Holism, Language Acquisition and Algebraic Logic". *Linguistics and Philosophy* 25, pp. 419–452.
Dresner, E. (2003). "Meaning Similarity and Semantic Vicinity". In *Human Reason: Papers in Honor of Marcelo Dascal*, ed. by Y. Senderovich and N. Zauderer. Tel Aviv: Tel Aviv University Press (Hebrew).
Dummett, M. (1991). *The Logical Basis of Metaphysics.* Cambridge, MA: Harvard University Press.
Ellis, B. (1968). *Basic Concepts of Measurement.* Cambridge: Cambridge University Press.
Fodor, E. and E. Leopre (1992). *Holism: A Shopper's Guide.* London: Blackwell.
Helmholtz, H. V. (1977). "Numbering and Measuring from an Epistemological Viewpoint". In H. V. Helmholtz, *Epistemological Writings.* Dordrecht : Reidel, pp. 70–108.

Holder, O. (1901). "Die Axiome der Quantitat und die Lehre vom Mass". *Berichte uber die Verhandlungen der konigliche sachsischen Akademie der Wissenschaften zu Leipzig Math,-Phys. Classe* 53, pp. 1–64.

Krantz, D., D. Luce, P. Suppes, and A. Tversky, (1971–1989). *Foundations of Measurement (Vol. 1-3)*. New York: Academic Press.

Mundy, B. (1987). "The Metaphysics of Quantity". *Philosophical Studies* 51, pp. 29–54.

Pagin, P. (1997). "Is compositionality compatible with holism?" *Mind & Language* 12, pp. 11–33.

Pagin, P. (2006). "Meaning holism". In *Handbook of Philosophy of Language*, ed. by E. Lepore and B. Smith. Oxford: Oxford University Press, pp. 213–32.

Quine, W. V. (1951). "Two Dogmas of Empiricism". *The Philosophical Review* 60, pp. 20–43.

Quine, W. V. (1974). *The Roots of Reference*. La Salle, IL: Open court.

Quine, W.V. (1986). "Reply to Gibson". In *The Philosophy of W. V. Quine*, ed. by L. Hahn and P. Schilpp. La Salle, IL: Open court, pp. 155–157.

Stevens, S. (1946). "On the Theory of Scales of Measurement". *Science* 103, pp. 667–680.

Swoyer, C. (1991), "Structural Representation and Surrogative Reasoning". *Synthese* 87, pp. 449–508.

5

Ornamentality in the New Media

ERAN GUTER

1 Introduction

The term "ornament" is commonly reserved for certain fixtures in our daily life such as Persian rugs or tacky wallpaper. For historical reasons, aestheticians have opted to downplay the philosophical import of ornamentality. Still a number of more recent leading lights, from Ernst Gombrich (Gombrich 1979) and Rudolf Arnheim to Kendall Walton (Walton 1990) and Peter Kivy (Kivy 1991), have become acutely aware of the cognitive value of ornamentality, and of the fact that ornamentality is an aesthetic phenomenon, which is much more widespread in art and in life than we tend to acknowledge. In this essay I pursue this line of contemporary thought as I offer some reasons in support of the seemingly strange claim that ornamentality is pervasive in the new media. I then turn to explore some of its ramifications, which yield, I shall argue, an interesting puzzle.

The term "new media" is commonly used as a blanket description for a whole range of different objects, processes, and practices, which have been growing increasingly intertwined and which already pertain almost equally to the domains of communication, entertainment, and lifestyle. What we ordinarily count as new media may consist of one or more of the following standard categories (Lister et al 2003: 13): (a) computer-mediated communications, primarily e-mail, chat rooms, voice image transmissions, the web, and mobile telephony; (b) digital technologies for distributing and consuming, primarily media texts characterized by interactivity and hypertext formats, such as the world-wide-web, CD-ROM, DVD, and the various platforms for computer games; (c) virtual reality, which runs the gamut from simulated environments to fully immersive representational spaces; and (d) a whole range of transformations and dislocations of established media in, for example, photography, television, film, etc.

There is a common temptation to adopt technological essentialism regarding the new media, that is, to identify the media with the technology. So we might say quite trivially that the new media are new because the technologies underlying them are new. To be sure, some of these technologies are indeed new and quite exciting. But others may not be so new, or are based on old ideas and conceptions. Consider, for instance, the notion of interactivity, which is widely held to be a constitutive characteristic of the new media. We ordinarily speak of new media interactivity quite literally, as consisting in physical interaction—real or simulated— between the user and his gadget: pressing a button, choosing a link, cutting, pasting, dragging an object, and so on. Yet, as Lev Manovich rightly pointed out, all classical as well as "old" modern media— literary and dramatic narratives, visual and three-dimensional representation, music, architecture, cinema, to adduce the most obvious examples— are interactive in the sense that they invite or hinge upon cognitive processes of filling-in, hypothesis formation, recall and identification, etc. (Manovich 2000: 55-61). In this sense, new media interactivity is not that different, and restricting ourselves to technological newness amounts to taking a one-sided view of a much richer picture—that of the enmeshment of our minds and lives in the technology. One should be reminded here of the prophetic words of new media pioneer Douglas Engelbart, advising his peers on the brink of the digital revolution to transcend technological essentialism: "We do not speak of isolated clever tricks that help in particular situations. We refer to a way of life in an integrated domain where hunches, cut-and-try, intangibles, and the human 'feel for a situation' usefully co-exist with powerful concepts, streamlined terminology and notation, sophisticated methods, and high-powered electronic aids" (Engelbart 1962: 1).

Thus, it would be safer, or at least it would make more contextualist sense, to say that the new media give rise to mediumal hybridity. According to Jerrold Levinson (Levinson 1990: 26-36), a hybrid is primarily a historical thing, a product of a certain development from concrete origins, which has emerged out of a field of previously existing activities and concerns, two or more of which it explicitly combines in some sense. We may take, for example, kinetic sculpture, such as a mobile installation by Alexander Calder. In Levinson's terminology, such an artifact is a "transformational hybrid" of two previously existing distinct art forms: sculpture and dance. We may say that it is a case of a sculpture gone dancing. It incorporates some of the special or distinctive characteristics of dance into what remains recognizably sculpture, albeit in an extended sense. Its mobility defies its solidity. Still what is significant about such a work of art transcends the technological achievement underlying its hybrid craftsmanship. It rather pertains to its entry into the realm of art, to the place that this artifact occupies in our lives, and to our response to its defiance of solidity by the suggestion of movement.

Now consider the example of digital TV. This new transformational hybrid brings together two "old" media: television technology—itself a direct descendent of the telegraph, the telephone, and the early analog technology of photo-telegraphy or facsimile—and computer technology. It is quite literally television gone digital. Yet what is new and significant about digital TV is not so much its technological hybridity *per se*, but rather the changes and effects that such hybridity brings about considering the whole environmental transaction pertaining to this medium, including what it can or does offer and what users do or can do with such offerings, and how this whole package is integrated into our living spaces and activities. In what follows I would like to focus exclusively on this sense of newness; hence I opt for a more "ecological" approach in this venture on the relatively uncharted terrain of the philosophical aesthetics of the new media.

2 Ornamentality as Inhibition of Games of Make-Believe

So why, how, and when are the new media ornamental? One answer, taken squarely from ordinary experience, readily suggests itself: at least some of these technologies are conducive to audio-visual styling; hence they serve a clear decorative purpose as fixtures in our daily life. Actually this has been the case with television—the immediate ancestor of most of the new media—almost since its very inception, as I have argued elsewhere (Guter 2002). The intertwining of such technologies as digital television, the internet, and mobile telephony decorates simply by virtue of contributing to and shaping one's environment in very much the same way that Persian rugs or flowery wallpaper do. The activated technology often becomes simply part of the space in which it is located; for some people, it may make an apartment homier and more inviting, perhaps even more than an antique sofa or a decorative chandelier. One may walk in and out of a room, catching no more than a glimpse of the screen, overhearing only a few words or a brief musical passage, and yet keep the technology switched on for pretty much the same reason one does not peel off and re-paste the wallpaper each time one exits and reenters a room. This point may be reinforced by observing the habitual frenzy of zapping and surfing. Such common practices often serve the clear decorative purpose of creating or adjusting one's ambience. Furthermore, at the very heart of contemporary art we find works which capitalize on the ornamentality of the new media, from the jocular early video installations by Nam Jun Paik to recent ontological conundrums of cutting-edge verve such as Tod Machover's *Brain Opera*, which utilizes the input of random anonymous internet users to shape immersive audio-visual environments in which real people roam, taking active part in the creative processes underlying this peculiarly shape-shifting work (Ippolito 1996).

Yet there are still deeper reasons for the claim that the new media are ornamental. Here I would like to refer to Kendall Walton's theory of ornamentation, which he presented in his influential book *Mimesis as Make-Believe* (Walton 1990). In his book, Walton articulates a general theory of make-believe, which is applicable far beyond the realm of figurative painting and sculpture, stories and novels, and other such artifacts, which exercise our powers of imagining. According to Walton, "propositions that are 'true in a fictional world,' or *fictional*, are propositions that, in a given social context, are to be imagined as true. What is to be imagined usually depends on features of real world... Things that generate fictional truths in this manner I call *props*. *Representations* are objects whose function or purpose in a given social group is to serve as props in games of make-believe" (Walton 1991: 380).

It is easy to see how Walton's theoretical framework can be deployed for our purpose here. Most of our new media experiences can be described quite straightforwardly in terms of using props in a variety of games of make-believe, perceptual or other, wherein such props can be, for instance, other network users (real or fake), texts, visual images, pop-ups and interactive graphics of all sorts, computer icons, navigational objects, sound effects, audio-visual clips, live feeds, and other stuff that new media dreams are made of. Our various games of make-believe with these props generate fictional truths about the props themselves, about the fictional worlds that they inhabit, and about us, the participants, or rather *users*. Furthermore, insofar as our new media experiences are mediated by man-computer interface, information takes the form of representation, whether by words, sounds, graphics, visuals, or, in certain immersive environments, even by kinesthetic sensations.

Walton offers an insightful account of ornamentation in terms of inhibition of participation in games of make-believe (Walton 1990: 274-289). Contrary to the standard case of fully-fledged pictorial representation, decorative representations are conceptually more complex in the sense that they present us with fictional worlds in which other fictional worlds are embedded. This puts us at a certain psychological "distance" from the embedded world, since we participate only in the first-order game of make-believe, while imagining that there is another game, which we could participate in. In Walton's words: "We stand apart from the internal fictional world and observe it through its frame" (Walton 1990: 284). For example, when I look at flowery wallpaper, I am withdrawn from an internal world, which contains certain flowers, into a second-order world of complex flowery patterns on the wall. Insofar as a representation is decorative, we inevitably find ourselves withdrawn to the point of being merely spectators, rather than participants in a fully blown game of make-believe. We oscillate between the tempting fictional richness of the internal world and the overpowering sparseness of the framing world, which consists of "scarcely more than the work itself together with, by implication, its artist and his creative activity" (Walton, 1990: 287). Importantly, Walton observes that this is also true of *bona fide* representations. Consider, for instance, Vincent Van Gogh's painting *Starry Night*. According to Walton, the physical properties of the painting—the bold brush strokes, the cracking of the paint, the swirling frenzy of the artist's pictorial language—pull us back from a particularly seductive internal world into a more "objective" perspective, which might indeed yield more significant connections with our lives (Walton 1990: 288-9). A clear advantage of Walton's account is in the way it shows how widespread ornamentality really is. It can be temporary or partial, coexisting with genuine representationality.

Considering that the new media are conducive to audio-visual styling, hence to decoration, we can see how this observation readily maps onto Walton's idea that ornamentality is to be explicated in terms of inhibition of participation in games of make-believe. For styling simply draws one's attention away from the represented object to the way the representation is actually produced, hence away from any fictional truth it may generate concerning the represented object. This is very obvious in the case of audio-visual styling in the new media. As I noted before, some new media technologies serve a clear decorative purpose. This is true both in real life and virtual reality environments, as well as in the interweaving of both realms, as the phenomenon of *Second Life* spectacularly shows. Furthermore, the very notion of "data-aesthetics" (Vesna 2007), that is, the creative quest of rendering mere information or patterns of data as being pregnant with aesthetic significance, has imbibed the basic truth, which Walton's theory of ornamentality fleshes out, that experiencing an object (a conglomeration of data) through and encased by its mediumal manifestation (man-computer interface) is tantamount to interfacing to a culture encoded in digital form (Manovich 2000). Thus, audio-visual styling in the new media yields significant connections with our lives by pulling us back to a game of "cultural interface".

Yet even in the realm of mere text, we find pervasive styling in the new media in the form of hypertextuality. Stated carefully enough in non-essentialist terms, this contention circumvents both the literary and the expressionist fallacies, which Luciano Floridi observes in the standard interpretation of hypertext (Floridi 1999: 121-123). While it is true that the idea of hypertext originated in an attempt to solve the problem of information overload by means of associative linkage of data (Bush 1945), it is also true that hypertextuality may yield in practice highly stylized texts such as John Cayley's celebrated self-assembling poetic collage *Book Unbound* or Stuart Moulthrop's technologically innovative essays and works of fiction. Such works of hypertext, whether consisting in passive link-node structures or automatically generated by an algorithm, are akin, in a sense, to some well-known modernist attempts in music—such as Karlheinz Stockhausen's *Klavierstück XI*—to generate "mobile form" by employing various types of chance operation, including associative linkage of precomposed fragments. We may safely say that insofar as hypertextual styling empowers the reader to determine the format of the text, thereby deflecting the reader back to the manner in which the text is generated by the reader's own performance of reading, it inhibits participation in games of make-believe. And if hypertext is indeed the conceptual structure of the infosphere—the ever expanding and converging digital "encyclopaedic macrocosm of data, information, ideas, knowledge, beliefs, codified experiences, memories, images, artistic interpretations and other mental creations" (Floridi 1999: 8)—as Floridi maintains (Floridi 1999: 128-131), then ornamentality is in effect the fundamental aesthetics of the infosphere.

Hypertextual navigation is a particular case of another key principle of the new media: interactivity. As I pointed out in my introduction, in the cognitive sense of interactivity—filling-in gaps, concluding, expecting, and so forth—the new media are not uniquely more interactive then any of the "older" media. However, contrary to the fixity and continuity of the analog media, digitality allows for random access to any data element, and it enables individual members of the new media audience to directly intervene in, and change, the data accessed. Such digital interactivity can be extractive, as in the case of hypertext, immersive, as in the case of virtual reality, or registrational, that is, consisting in writing back onto a database, as in the case of internet bulletin boards and multi-user domains. Either way, digital interactivity amounts to world-building—simply put, the viewer becomes a user—which means that when we digitally interact with the medium we patently refer back to the features of the medium itself, we are withdrawn to the way the representation is actually produced. In this sense we may conclude, perhaps somewhat contra-intuitively, that digital interactivity in general inhibits participation in games of make-believe. Furthermore, as David Z. Saltz pointed out, such interactions are performative "when the interaction itself becomes an aesthetic object; in other words, ... to the extent that they are about their own interactions" (Saltz 1997: 123). Thus, performativity is the semantic hallmark of the user's participation in what Walton's theory refers to as the first-order game of make-believe, consisting of the user's own world-building activity, while imagining that there is another, embedded game, which he could participate in, albeit not as a user *per se*, i.e. as a "world-builder".

3 New Media as Conduits of Real Life

The new media have emerged from an unlikely hybrid of military oriented and aerospace technological drive, the need of the entertainment industry to employ the full sensory array of its audience and to accomplish new forms of narrative, the quest of computer and artificial intelligence research for a fully fledged man-computer symbiosis, and fancies that have been brewed in the deep recesses of cyberpunk subculture. Still the overarching identity of the new media as conduits of real life, that is, as means of communication in the broadest human sense, not just in the narrow sense of data transmission, was already evident in some of the early conceptual frameworks, which shaped the emergence of the new media, and it certainly became a fact of reality at least since Douglas Engelbart's groundbreaking demonstration of his online system in 1968. As conduits of real life—whether by means of text, image, or sound, as play or as work, amid our most ordinary routines or altogether in a virtual reality environment—the new media are most commonly accessed and interacted with in search of knowledge.

It is crucial to observe here that the digital medium in itself, in its technologically essentialist sense—the closed circuit, the technology merely being on—is characterized by epistemic transparency. That is, the digital medium is capable of presenting perceptual information that is caused by and counterfactually dependent upon its subjects (Walton 1984: 246-77). This is due both to the specific hybrid origins of the new media in earlier technologies of distant seeing and facsimile, and to the nature of digitization—the process of converting continuous perceptual information into a numerical representation by means of sampling and quantifying. The technology in itself has been designed to be absolutely inert with regards to the content, which it channels. Indeed we tend to perceive the many idiosyncrasies of the medium—such as electronic distortions, blurring or unnatural coloring of the image, which are rampant nowadays in video transmissions carried by third-generation mobile telephony—as having no bearing on the status of events and objects in the world. As a converse example, consider the medium of Western tonal music, which imposes clear structural requirements for tonal movement to occur. If similar requirements were to be present in a typical new media apparatus—for example, if a given segment of web camera feed were to have its perceptual equivalent of a perfect cadence—the perceptual information embodied in the segment would have been, at best, only partially caused by its subjects, and, in any case, it would not have been counterfactually dependent upon its subjects. Interestingly, John Cage, the visionary American composer, noticed this crucial difference already with regards to the "old" medium of television. Seeking to free music from the "totalitarianism" of Western tradition, marshaled by his onetime teacher Arnold Schoenberg, Cage preached for raising music to the condition of television (Cage 1961: 40).

These considerations suggest the philosophical significance of any introduction of boundaries into the clear medium, of the interactive compromising of the open channel. A prime example is simply the framing or cropping of the photographic image. After all, the real life channeled by the new media is a framed real life, truncated by the technical specifications of the equipment used and set to fit our gadgets. As Stanely Cavell pointed out, the significance of the photographic frame lies in the brute fact that the photograph comes to an end. "When a photograph is cropped, the rest of the world, and its explicit rejection, are as essential in the experience of a photograph as what it explicitly presents" (Cavell 1979: 24). That is, the frame has a meaning internally related to the meaning of the image it encloses.

It may be instructive to recast this idea using the valuable terminological distinction, employed by R. M. Hare (Hare 1970), between the *phrastic* and the *neustic* of an utterance. By the phrastic, Hare simply means the propositional content of the utterance. The neustic is what Hare calls a sign of subscription to the speech act that is being performed: it is that part of the sentence which expresses the speaker's commitment to the factuality, desiderability, etc., of the propositional content conveyed by the phrastic (Lyons 1977, vol. 2: 749-751). This distinction is easily carried over to the analysis of pictorial or even sonic representation (although for reasons, which remain beyond the scope of this essay, it may not be readily applicable to the analysis of musical representation). Arthur Danto (Danto 1995) suggested that the phrastic of a pictorial representation—a painting or a photographic image—is what we normally take to be its propositional content: a tree, a man, the half-dome in Yosemite Park in California. The neustic of a picture would be the attitude its "author"—normally the painter or the photographer—wanted us to take toward that content: a certain feeling, a moral attitude, but also a commitment to its factuality.

Now returning to Cavell's insight, we can say that inasmuch as the frame puts us in some kind of relationship to the phrastic content of the photograph, it performs a neustic function. It enfolds and engulfs not so much the photograph as us, the spectators, together with what the photograph shows. My upshot is this: mediumal elements, which eventually deflect us back to the features of the actual representation, inhibiting our participation in games of make-believe with its phrastic content (our ability to generated fictional truths about that content), perform a neustic function. Thus, closing the circle, which began with my previous discussion of Walton's theory, I submit that ornamentality hinges upon the neustic. In fact, I suggest that this is actually what Kendall Walton meant by saying that ornamental representations pull us back to a more "objective" perspective, which might yield more significant connections with our lives (Walton 1990: 288-9). Furthermore, I conclude that the status of the new media as conduits of real life is intrinsically related to the neustic function of their mediumal idiosyncrasies. It is noteworthy that Walton's anti-realist, constructivist theoretical framework becomes a genuine asset here, since it enables us to construe the difference between real life and fictional representations in terms of different games of make-believe, which in principle might even employ the same props.

4 Ornamental Erosion of Real Life

Now, as Hamlet says, "there's the rub". If the new media are ornamental, then, insofar as they serve as conduits of real life, they are ornamental in a sense, which is very different from the case of flowery wallpaper or Persian rugs. New media ornamentality uniquely exemplifies ornamentality without abstraction. A pinkish wallpaper flower may be an abstraction of a particular flower, exemplifying all flowers of its kind, yet none in particular. On the other hand, the new media, insofar as they are used as conduits of real life, are all about particular things: names, faces, gestures, and events. Granted, we can now put Walton's theory of ornamentation to an interesting use. If we understand ornamental representations in terms of game worlds in which other fictional worlds are embedded, and if ornamentality consists in being pulled back to the more "objective" frame-world, then new media representations confront us with a puzzle: their internal worlds are inhabited by real life denizens, which become somehow "less real" by virtue of our withdrawal into a more "objective" perspective. This amounts to an ornamental erosion of real life.

The new media present us with real life cased with a distancing neustic frame-world that sustains a manifold of mediumal devices, some essential, like interactivity and hypertextuality, while others purely decorative and evocative, like audio-visual and graphic effects. Unfolding in time and spread out graphically in virtual space, bits of mimetic material, plucked from the flux of life, are set in elaborate, dazzling designs, like precious stones set in a glittering piece of jewelry. The result—kaleidoscopic, audio-visually stimulating, and seductive in many ways—leaves us oscillating tentatively between the fictional and the real. Insofar as we use the new media as conduits of real life and as means for knowledge-seeking, the excessive density of what I referred to as the distancing neustic frame-world—especially in such cases as internet-based virtual worlds or massively multiplayer online role-playing games—forces us to conduct our moves under conditions of neustic ambiguity, that is, uncertainty concerning the kind of relationship we, the users, have to the propositional content mediated. This seems patently true, at least in certain new media environments, if we define the neustic of a representation, as Danto suggested, in terms of the attitude the "author" of the representation wanted us to take toward its propositional content, since, as I already observed, the notion of a "user"—in the context of hypertext, hypermedia, and digital interactivity in general—undercuts, or at least problematizes the notion of an "author".

We may say that new media users operate behind an ornamental "veil of ignorance", yet in a sense importantly different from the one John Rawls had conceived for his purposes (Rawls 1971). Whereas Rawls's original "veil of ignorance" assumes ignorance of particular real life situations, the condition of new media ornamentality leaves them intact—carefully selected or utterly made-up—to serve as an opening move in a game of knowledge-seeking (Hintikka 2007). Yet the very nature of the game—some of its definitory rules, its goals and desired strategies—are bound to become ambiguous, if the inquirer's attitude toward his sources turns out to be ambiguous as well. This is clearly the case in new media environments such as *Second Life*, for instance, which capitalize on the extreme malleability of data by users, and their ability to fabricate immersive, intelligent environments by digital means. Within such new media environments, which are typically inhabited by various software applications designed to emulate human interaction, and which commonly involve intense role-playing, the self-identity of the user is patently rendered ambiguous. This point has been underscored forcefully, albeit with a distinct Lacanian bent, by Sherry Turkle: "In my computer-mediated worlds, the self is multiple, fluid, and constituted in interaction with machine connections; it is made and transformed by language" (Turkle 1995: 15). In other words, virtual identity is ornamental.

5 Conclusion

In this essay I contended that insofar as we consider the new media ecologically, not just technologically, the new media are ornamental. I located ornamentality both in the logically constitutive principles of the new media (hypertextuality and interactivity) and in their multifarious cultural embodiments (decoration as cultural interface), and I identified it as the ground-floor aesthetics of the infosphere. Considering how heterogeneous, polymorphic, and dynamic new media phenomena are, this might seem like a theoretical long-shot. Ornamentality is clearly more rampant in certain new media environments and practices than in others. Still, as Floridi observed (Floridi 1999: 14-15), the infosphere has been gradually evolving since the 1950s along three fundamental vectors: (a) toward multimedia information and virtual reality; (b) toward graphic and immersive interfaces; (c) toward integration and convergence of the global network. This entire process is conducive to new media ornamentality, and if Floridi's analysis is correct, then it is reasonable to expect that the various manifestations of new media ornamentality are bound to become more pronounced.

The centerpiece of my argument is the puzzle of the ornamental erosion of real life in the new media. This puzzle calls our attention to a peculiar interrogatory complexity inherent in any game of knowledge-seeking conducted across the infosphere, which is not restricted to the simplest form of data retrieval, especially in mixed-reality environments and when the knowledge sought is embodied mimetically. The puzzle is set up by means of Kendall Walton's theory of ornamentation. In this context, Walton's theoretical framework has a clear heuristic value, as it opens up possibilities for a sober analysis of the peculiar semantic complexity, which characterizes at least some of our engagements with new media. On the one hand, it saves us from falling prey to certain McLuhanian sentiments (McLuhan 1965) by reminding us that performativity does not necessarily imply a conflation of the medium and its message. On the other hand, it saves us from giving in to the postmodern urge to dismantle and dissolve the classical tripartite definition of knowledge as justified true belief. It should be noted here that the ornamental erosion of real life does not give rise to anything like Jean Baudrillard's philosophically extravagant idea of the successive phases of the image from being a reflection of reality to being its own pure simulacrum (Baudrillard 1988: 166-184). If my way of construing the puzzle by means of Walton's theory of ornamentation is viable, then there can be no slippery slope to the effect of a complete ornamental erosion of real life. In other words, given that the difference between the embedded fictional world and its frame-world can be couched also in terms of the distinction between being a spectator of a game and participating in one, there can be no conceptual room for an apotheosis of the mimetic within dense ornamentality.

The puzzle of the ornamental erosion of real life poses an interesting and rather unusual challenge for aesthetic cognitivism: to figure out what would be a viable logic of virtual discovery under the conditions of new media ornamentality. At any rate this must be an epistemology that focuses not on the classic project of justifying knowledge already acquired, but rather on how knowledge is acquired in the first place in new media environments; and here, as I have suggested, aesthetic concerns play an enormously important role.

References

Baudrillard, Jean (1988). *Selected Writings*, ed. by Mark Poster. Palo Alto, CA: Stanford University Press.
Bush, Vannevar (1945). "As We May Think". *The Atlantic Monthly* 176 (1), pp. 101-108.
Cage, John (1961). *Silence: Lectures and Writings*. Middletown, CT: Wesleyan University Press.
Cavell, Stanley (1979). *The World Viewed: Reflections on the Ontology of Film.* Cambridge, MA: Harvard University Press.
Danto, Arthur C. (1995). *Marcia King and the Symbolic Language of Frames.* East Hampton, NY: Guild Hall Museum.
Engelbart, Douglas C. (1962). "Augmenting Human Intellect: A Conceptual Framework", Summary Report AFOSR-3233. Menlo Park, CA: Stanford Research Institute.
Floridi, Luciano (1999). *Philosophy and Computing: An Introduction.* London: Routledge.
Gombrich, Ernst H. (1979). *The Sense of Order: A Study in the Psychology of Decorative Art.* Ithaca, NY: Cornell University Press.
Guter, Eran (2002). "Anti-Mimesis Live". In *Television: Aesthetic Reflections*, ed. by Ruth Lorand. New York: Peter Lang, pp. 139-160.
Hare, Richard Mervyn (1970). *The Language of Morals.* London: Oxford University Press.
Hintikka, Jaakko (2007). *Socratic Epistemology: Explorations of Knowledge-Seeking by Questioning.* Cambridge: Cambridge University Press.
Ippolito, Jon (1996). "Whose Opera is This, Anyway?" *Postmodern Culture* 7 (1). <http://muse.jhu.edu/journals/postmodern_culture/v007/7.1r_ippolito.html>
Kivy, Peter (1991). "Is Music an Art?" *Journal of Philosophy* 88, pp. 544-554.
Levinson, Jerrold (1990). *Music, Art, and Metaphysics: Essays in Philosophical Aesthetics*. Ithaca, NY: Cornell University Press.
Lester, Martin [et al] (2003). *Mew Media: A Critical Introduction.* London: Routledge.

Lyons, John (1997). *Semantics*, 2 vols. Cambridge: Cambridge University Press.
Manovich, Lev (2000). *The Language of New Media*. Cambridge, MA: MIT Press.
McLuhan, Marshall (1996). *Understanding Media: The Extensions of Man*. New York: McGraw-Hill.
Rawls, John (1971). *A Theory of Justice*. Cambridge, MA: Harvard University Press.
Saltz, David Z. (1997). "The Art of Interaction: Interactivity, Performativity, and Computers". *Journal of Aesthetics and Art Criticism* 55 (2), pp. 117-127.
Turkle, Sherry (1995). *Life on the Screen: Identity in the Age of the Internet*. New York: Touchstone.
Vesna, Victoria (2007). *Database Aesthetics: Art in the Age of Information Overflow*. Minneapolis, MN: University of Minnesota Press.
Walton, Kendall L. (1984). "Transparent Pictures". *Critical Inquiry* 11, pp. 246-277.
Walton, Kendall L. (1990). *Mimesis as Make-Believe: On the Foundations of the Representational Arts*. Cambridge, MA: Harvard University Press.
Walton, Kendall L. (1991). "Précis of Mimesis as Make-Believe: On the Foundations of the Representational Arts". *Philosophy and Phenomenological Research* 51 (2), pp. 379-382.

6

The Imaginary Pieces
AMNON WOLMAN

1 Two Descriptions

March 1998

I started writing the imaginary pieces in 1998.

They are a long-reach outcome of a debate I had with John Cage the final time I consulted with him in Evanston (in 1992) when I performed *Europera5* with Yvar Mikhashoff. At the time one of our students asked John the (transparent) question: *What is music?* And John responded (after the well-versed theatrical pause and the small winning smile): *It must have something to do with sound.*

I (erroneously) postulated that he meant *that it must have something to do with physical sound.* And I interrogated him about illusory timbres. Sounds that I detect when I conceive of them. These are compositions that are actualized in my mind, and have no province in the material sphere. This led me to an abundance of investigations with utterances that evoke sounds and time. Consequently, I comprehended that the recollection of music does not tackle continuance, as it seems to be perceived in the physical realm. It shrivels time towards an individual instant. For example, when a person pronounces the words "Bach Goldberg's Variations" I respond with a clear acoustical memory, a sound in my mind, which is an encapsulation of the whole piece with the passage of time embedded in it. It is not just the label of the name of the piece, but an allusion that embraces time and the music's advancement in it. I attempted shaping such moments in my music, individual instants that were designed to elicit the continuousness of time. Recent pieces present an effort to invent a composition that in each and every moment depicts its whole.

This year I adopted a changed route, fabricating pieces that transpire only in the audience's mind. They have no physical attributes. Each has disparate explicit directions of how it should be presented: Some are executed while other music is being played and others in silence. A piece is invariably performed by providing every member of the audience with a reproduction of the score, to be read in silence without an origination of any physical sounds.

When writing these pieces I unraveled new things; about time propelling backwards (it is not symmetrical to the motion forward), and about the perceived and determined beauty of sounds, and most of all about our agreements as audience members about what it is that we listened to.

A New Description – February 15, 2009

The interest in text pieces came about from two directions. I started writing text scores for improvisation ensembles in 1988 when working on *Nautilus*, a collaboration of live dance/video/music at San Francisco State University. These pieces use texts to convey instructions to musicians, and are made of language taken from the traditional discourse of western culture musicians. The most useful feature that came out of this pieces/process/research was that I found out that I could specify time in the score in a non-linear way. For example, consider the instruction "start by playing this melody and continue improvising on it on and off for the next few minutes; when you are not improvising you may whistle a tune or dance a jig." (This is not a quote from a specific piece but a useful example.) In this example the discourse is one of western musicians— "melody", "improvising", "tune", "jig" are all words that have specific meanings in this discourse. But Time, in this example, is flexible, i.e., the composer doesn't know how long this section of the piece would be ("few minutes") and also non-linear. The text describes the action for the full length of the section ("on and off for a few minutes") and then goes back in time to fill, with new sounds or actions, the holes that were left in time.

The second direction, which led me to a new and different type of text pieces, is based on a response that John Cage made in one of our classes at Northwestern University to the question "what is music?" To which he responded, "it must have something to do with sound." It took me several years to ponder this statement, and discover that what made me uncomfortable about it was my interpretation that sound is a literal physical entity; but the other aspect of "sound", the imaginary one, the one we can recall in our mind without any physical sensation—I treated that as a separate entity. This started a process of compositional activity that I group under the heading "imaginary pieces". The listener in her/his mind, without any physical sensation, creates these pieces. The discourse here uses the English language within its normal day-to-day use, and, importantly, within the bounds of western culture. So images like "the beeping sound of a life support system" are used to evoke a sound, which, it is assumed, is familiar to any TV watching western culture based audience and is quite precise in its sound contents. And when the image "a flute like pretty melody" is used, it is intended to evoke a less determined sound but one about whose sound contour we, participants in western culture, know many things. Composing in the mind of the audience entails, I believe, describing the sounds in words, so the choice of language and the determination of its cultural context are imperative. And, obviously, this encourages the translations of the text to the language of the locale of the performance. Since I believe that most performances of the pieces happen on-line, English seemed to be the most appropriate and relevant choice.

The term "discourse", rather then "performance practice", is chosen because "discourse" within our academic/artistic discourse clearly identifies cultural context and social structures as components of the discussion; on the other hand, the term "performance practice" is placed within musicians' discourse and seems to make these two distinctions blurry. "Performance practice" refers to a body of knowledge that a musician performing a piece uses. "Discourse" encompasses both that discussion and the body of knowledge that the audience employs when it listens to a piece of music or is involved in the performance of music.

I found two very powerful tools in this discourse. The first is that it provided me with the ability to mix ugly sounds with beautiful ones without having the "ugliness" become an issue of discussion between the composer and the audience. If asked to imagine "ugly" sounds, an audience member will choose a sound that is appropriately ugly for her/him.

The second tool is derived from the ability to choose words according to how they sound and place them next to each other to create rhythmic patterns; this is obvious and elementary in the discourse of poets. "This sound is tremendously slow, tremendously slow," attempts to convey the sound by using the sound of the "s" and makes an effort to influence the rhythm by repeating the words. This, as in poetry, may get lost in a translation, so the use of it is limited.

In the imaginary pieces the audience is provided with a sheet of paper on which the text is printed, and given a certain time to read the text and play the sound in their head. The notation and layout are important as they do imply certain musical elements. Size of words and the distance between them are, it is assumed, translated into accents and spaces in time in the imaginary music. There was a suggestion that I project the text one line at a time, gaining more control of the time process, and the unfolding of the piece in the mind. I may do a piece like that in the future (perhaps very soon) but I found that idea problematic for the older pieces, significantly because the flexibility of non-linear time may be lost. For example, a sentence such as "earlier a melody sung by an old female Arab voice started but it just now became noticeable," may lose the power it has of moving backwards in time. If time is explicit as it is when the line appears in the projection in a pre-specified moment, the conviction that we have moved backwards in time may be lost. But, as I said, this new idea has my interest, and it does seem to solve one problem that the text pieces have: that descriptions of time of different durations are read at equal duration. For example, "the low timbre becomes louder very slowly," and "the high sound becomes softer very quickly," take the same amount of time to read, but should take different lengths of time to play in one's head.

As I am writing this I am reading Oliver Sacks' book, *Musicophilia*. It's a wonderful, amazing book that summarizes the current research on how the brain processes music, sound, and the memory of music. There is abundant evidence that the ability to imagine music and sound is very common to all of us, and the ability to create new music in the mind, one that is based on memories of music, is also common. My imaginary pieces, in this context, seem very obvious, and not at all radical or unusual.

The imaginary pieces present an expression of another one of my interests, that of audience/performer/composer interactions. For many years I have struggled with the role of the audience and its placement in the physical world of music representation, and how it influences the music itself. There are many different issues that arise in this discussion. One of the first that comes to my mind is that of the traditional concert setting, where the audience is comprised of a trapped group of like-minded individuals attached to their seats for the duration of the music. In any live music concert the audience is identified as a group of equal members, equal in that they are all there for the same reason—to listen to the music. This can be, and is, sometime translated to mean that they are also going through an identical, and mutual, experience, and that everyone hears the music in the same way. This is the basis of, for example, music criticism, where the critic is responding to a performance, which, it is assumed, all attending heard in the same fashion. This is a notion that I reject: many times I have come out of a concert and, while attempting to describe to my friends what it is I heard, I have discovered that they heard something else altogether. In the classical music world the situation is even more restrictive: the audience is placed in such a way (with the lights on) that leaving the hall in the middle of a piece becomes close to impossible. Obviously it is a choice that they made responsibly, in advance, but can people not change their minds?

But perhaps the most significant question that arises, for me, as an aspect of the traditional placement of the audience, is that the sound we hear as musicians is so different from the one the audience hears. When one is on stage, for example in the orchestra, the choir, or next to a single pianist, the sound one hears is much richer (to my mind) and all engulfing than any sound projected to the audience sitting in the auditorium and hearing the music, either directly or through audio speakers. How do you get the audience to move among the performers, to feel free to move in and out of the space, and to get a sense that their experience is both individual and collective—these are questions that I have tried to solve, to varying success, in different pieces (*The Andy Warhol Diaries, NU Piece, Cruising Prohibited When Lights Flashing*), in installations (*the Speakers Suit, The Speakers Army Jacket, Attention Step*), and in the imaginary pieces. In the imaginary pieces the individual and private is expressed and emphasized, the audience member is the performer. The total experience and the sounds created to represent it are clearly individual, and it is the commonality of the experience that is uncertain and may be discussed. The freedom of the choice of involvement with the piece is held exclusively within the prerogative of the listener/performer.

2 Imaginary Pieces 1998-2001

> A noise. A dainty, vintage noise.

February, 1999

> In our mind's ear we start an electronic piece, performed on eight audio speakers that are placed surrounding us, in a space, which is in near to total blackness. It is so dark that we can sense the person sitting next to us, but we cannot quite see him or her.
> It starts with rumbling sounds that moves fairly fast on the right. The sounds emerge from total quietness to a comfortable level of loudness, and just stay there. They keep changing and yet nothing seems to happen. Suddenly the sounds stop and there is a sense of absolute silence. We hear breathing and are not sure if it is someone sitting next to us or emanating from the speakers. The same rumble that was heard in the beginning appears again sounding very very softly at the back of the imaginary performance space and after a certain time

disappears. Sounds of breathing and very muted talking are heard as if a person was walking above us on the roof. Its not clear if they are coming from outside the physical space we are in or from the audio speakers. The initial rumble appears again, and it's getting louder, while it does that it also moves from the left corner of the room to the center and transforms itself into a single pitch that sounds like a cross between trombone and violin. The pitch stops but a little echo lingers, meanwhile you notice that sometime before that a different sound of regular breathing has started. It could be an animal breathing and it has been there for a while, it is accompanied by sounds of muted drums. The breathing is rhythmical but does not match the rhythm of the drums. The breathing is so rhythmical that it seems unnatural. The drumming or hitting sounds remind us of strange tribal drumming. The single pitch on the trombone/violin appears again and it is accompanied (or creates) a chord. A beautiful chord, you have never heard this chord before, it is suspended in the middle of the space, and it sounds turn into voices sung as if through plastic tubes. All of the notes but one drop off and this last one lingers for a long while. It is a central note from the chord not the highest or the lowest. The other voices dropped off one by one, dissolving not together or abruptly. The single sound moves around the room getting faster and quieter, at the last moment, when it is almost inaudible, it glissandos down to silence. Suddenly we realize that the drums have now become a regular drum groove, pretty fast, and dance like. There are no changes in pitch areas but a clear sense of several drummers. The sounds then move from far away in the distance closer and closer almost to our chest. A sound of a strange bird crosses above followed, after a short silence by the sounds of a group of birds of different types. The sounds of these birds mask all sounds, when they leave we realize that the drum-sounds were replaced by a rumble, this time a lower one, and slower one. It has, sometimes, very high pitch properties, which sounds as if it is shimmering in the dark. Very far away at the distance, slightly to the right, we hear very soft breathing alongside another beautiful chord. The rumble disappears slowly. One more "bird" moves to the center from the back and then back from where it came.

Silence, it is not clear if the piece has ended, and if the breathing we hear is ours or sounds played through the speakers.

November 4, 1998

A loud vibration that is not explicit. We don't know if it represents a howl or a bomb, a gong or a plane. It has abundance of bass sounds that are felt in your stomach more then heard, and a high end that is piercing and ends very quickly. It is ensued by sounds of people talking. We just hear them pronouncing, not to each other, just vocalizing. It is the sound of talking, and this is music, there is no meaning to what they say. We just hear sounds.

October 12, 2000

A solitary and pendulous drone that appears akin to bead of water at the end of a stalk on a very soggy day. We sit listening and observing, expecting it to end and break off. And it doesn't and before long we stop waiting and go back to our train of thoughts. When we listen again and "look up" the sound, the drop, has already landed it is no longer there.

July 12, 2000

In this sound-image we are walking on the beach, maybe in Cannes France (where I've never been.) The resonance of the breaking waves is constant, but we don't always hear it. When our fascination is taken elsewhere the sounds of the waves vanish. Now its here and we hear it, and now it disintegrates. We hear the sounds of the birds, Seagulls I think, in much the same way; every once in a while a screech makes us focus our observation and notice that the birds are calling. People walk all around us, some going in the opposite direction. We listen to the sound of a language, it is mostly French sounding and we listen to the timbre of the language not the content. It is a color coherent to others but to us it is just an exquisite noise. There is a lot of salt in the air, we smell/sense it. The chatter around us suddenly sounds ominous, what are they saying? Should we try to get to the meaning of words and let the meaning of sound become less important? When those people walking in the opposite direction get closer we hear their voices more distinctly, but they disappear fairly hastily. There are also people playing ball on the waterfront and the sounds of hitting bodies is more sensed or fabricated then heard. Among the people walking in the same direction, some stroll slower then us and others stride faster but we can hear more of their conversations. A woman making a speech in a high nasal voice, and a kid repeating a question or is it a statement? A father sounds as if he is lecturing a young woman. The dialogues are distinct and do not blend into each other, they are only broken by the sounds of the waves. We hear a tune that sounds familiar, it is incomplete, it may be something we know and yet we can't place. It never terminates or goes anywhere it just starts. We hear the humming of it - are we murmuring along with it? Brigitte Bardot says "Poire," "poo" (or "po") "are." The sound of the "r" is ever so soft.

June 13, 1999

La la la la la la.
La la la la la la.
La la la la la la la la la la la.

La la la la la la.
Tra la la la la.
La la la la la la la la la la.

La la la la la la.
La la la la la la.
La la la la la la la la la la la.

La la la la la la la la la la la la   la la la la la la la la la la la.

February 20, 1999 (for Ruru)

A long, elastic sound, not too high yet one that seems to be growing in time. We are not sure if its growing in volume or in pitch, but suddenly we become aware that another very quiet, scratchy sound is in the background. It has been there for a while. It all stops and we are left with the sound of rain.

April 18, 1999

The first sound we hear is the sound of a drop of water falling on a hallow piece of wood. It repeats in a regular pattern and then suddenly it stops, followed by silence. It is an amazing complete silence. We choose to ignore and not follow other sounds, which are present, we just get the sense of a silence, quietness, and evenness. I was in the crowd watching the Gay Parade in New York in 1998, when at one point there was an official moment of stillness as a memorial to the people who died of AIDS. Everyone around us stopped and was silent, the throng watching, and the crowd and floats parading, were all silent. It was the sensation of silence that snared me. It was not a physical silence and you could not hear the blood in your veins, but you could feel it. This is what drives the soundlessness in this piece. It is what we hear now. It is a silence that is pliable. We can change it by sometimes hearing the sounds outside and at other times choosing not to detect anything. But suddenly a loud sound appears, it is abrupt but does not make us jump, this is because it had a small amount of increase built into it. This loud sound is followed by a very long and rich reverberation, which is somewhat filtered. In this decaying reverberation we can almost hear people singing or holding pitches. These almost voices stay for a very long time. No additional information is provided, new sounds do not appear, and the voices are slowly receding.

June 30, 1999

A new sound, a very nervous, taut and nerve-racking sound. You hear it and almost want to put your finger in your ear to dampen it, but you hold back, reasoning that it will mutate, it will turn into something else, but it doesn't. It is an annoying piercing shrill, and like a "difference tone" it is seemingly placed inside your ear. It persists and perseveres. It is not quite electronic, or not what we think of when we say an electronic sound. It clearly has a source, possibly it originates from a machine, a drill that someone controls with their finger and soon they will stop, but they don't. It just keeps on going. Every once in a while the volume seems to have a bearing and it is possible that perhaps it will disappear, but it doesn't. It just keeps on going. It continues and continues – persisting on and on. When we spin our heads it seems to alter only slightly, but then you recognize that it didn't really change. It is not clear if that change was in our head, the result of wishful thinking or an actual transition. But it gets softer now, it may even end, but it doesn't. It just keeps on going. Suddenly we are aware that we have no power over it, we can not turn it off, we can not leave, it will stay until someone else will turn it off, until the composer (shall we call her/him that?) will decide to move to something else, another musical gesture. Thinking about such a notion, the prospect brings us some relief, but our focus is brought back to this annoying constant buzz. It seems that perhaps it is not always the same. It sounds automated like a the noise made by a tool and it is not repeating, so when we succeed in thinking of something else it makes these minute small changes and our attention is drawn back to it. It is still there all the time.

February 26, 2000

A conglomerate of sound perpetually altering, never changing. It appears and sounds like a relief, an embossment; one can nearly trace it with one's fingers. It undulates and billows with a rhythm of its own that is difficult to disentangle and yet suggests regularity, an organic melting with the sounds of our bodies. The noise continues and we relent to it. While it shimmers at the edges, and proceeds forward, it indisputably never budges. There is an instant when we resolve or grasp, actively, to linger within it, not get bored, and its inmost vigorousness is there, pliant, yet never our own. It is a timbre outside ourselves that we will never master in penetrating. It will invariably remain a mystery.

When it terminates, we feel abandoned, dispirited, and that a protracted length of time passed. We have not absorbed it, but it mastered a circumstance connected with us that we are uncertain as to what it is.

And now we are left with soundlessness, the soothing noises on which we don't focus. We leave.

October 27, 1999

A very drawn out and dragging bass sound, it starts exquisitely and gets louder, but never really gets booming.
If you could you would turn it up just a bit.
In it you hear further tones, perhaps the sounds of persons singing or talking, but you are not confident that they are indeed there, maybe you are just conceiving them mentally. It suggests an entity intensely basic, and yet you (or I) have no idea what "basic" means, it just seems like a precious assertion used to specify that sound. It ends abruptly, into noiselessness, artificial silence, one that makes you very aware of the sounds that the lights in the room make.
The silence is so long, you speculate that this is all, that the piece is complete.
But then once again the bass sound sets forth. It appears like the one we listened to initially and you believe you understand all there is to appreciate about it. You almost tune it out; "there is no need to listen to this, I know it, I can think of something else" but abruptly you are conscious of a timbre that you felt or heard. Someone is saying something. It was nearly audible and you are convinced you did not conjecture it, it was there, and you are waiting to hear its next occurrence, but instead you detect in the vista a long melody played on some mysterious string instrument (violin?). It is a folk-like tune, intensely lengthy and slow winded, and you recognize that adjacent to it, there is a new bass that truly describes an unusual modal sequence of harmonies. It is a folk song of a tribe you are confident does not exist, and you do not understand how you are so undoubting. Abruptly you realize that the low-pitched sonority that launched the music disappeared a while back, and you are not certain when or how. The melody vanishes into an intensely noisy silence, as if every member of an orchestra was rubbing his/her hands together. It stops; it suddenly feels very lonely and cold.

March 26, 2000

We hear a squeak, scarcely distinguishable, as if someone is opening a very diminutive gate and unexpectedly resting. There is a protracted silence subsequently, the kind of cessation of sound in which you hear your own breathing louder then the noise of the traffic a ways off. Your breathing attenuates; it is a sound of murkiness and you are not certain if your eyes are open or closed. Very softly you hear someone humming; it resonates like a folk tune of a culture you may belong to but not resolutely. Perhaps you just think you know the person who is humming. This purring-like sound disappears leaving a very clear buzz similar to the sound made by a very large swarm of flies. All of a sudden, but without a dramatic gesture, traditional music begins, it is a four-voice piece of music played on what sounds like string instruments. The music is very languid and mellow, made of long phrases followed by very drawn-out pauses. It is music that sounds always similar yet never repeats. The loudness is steady, and the music rarely changes its sense of motion it just plays on. Casually the string-like sounds and music ceases and we are left with a single suspended note held for a very long time. We anticipate that it will end, then we want it to end, it stays on for too long, and then we live with it. It will continue until stopping and we listen to the tiny minute fluctuations within it.

June 22, 2000

The last vibrations of a friend inhaling and exhaling his sighs. It's a resounding sound comparable to a wheeze but emanating from his mouth. It is climactic and dramatic, full of exertion yet effortless. Time dissolves. Every breath is a composition to itself. The span and space between an intake and the outbursts stretch further. We anticipate the exhale with unease, and at each juncture when it is attained it is a comfort. The sound is not earsplitting yet it is the most deafening one we've ever heard; it is not a scream. Don't make any mistake, you will find no exasperation here only peace and reconciliation. It is life. All other sounds, the drip machine, the oxygen, the TV in the next room, the nurse arguing with another family, all are irrelevant and unrelated. They do not serve as an accompaniment to it; they are just there and have nothing to do with our sound. With our whispers, and the noise of a rearranged tissue paper, we all partake in our sound. We all inhale together, linger, and at long last set the air free. This sound of ours lasts forever, endlessly. And it stops.

For Ricardo, who died of AIDS on December 28, 2000 at 11:20 PM, with love.

January 4, 2001

A simplistic piece constructed on one idea presenting one simple motion. It is based on sound that we all understand, (like a baby crying.) It is manipulative in its audio single mildness. A piece I would not like to create.

May 19, 1999

The sound of a car far away. It gets closer and moves away. As it leaves we hear a single low drone, very low; it feels physical, as if it emanates from our insides. It keeps changing - expanding and contracting. It moves with us. At times there is almost a sense as if it is breaking up into what sounds like a rhythmic entity similar to low drums or heartbeats. There was another fleeting sound very high, a sound similar to a seagull but more synthesized that encompasses within it a multitude of after effects, and when it is over it leaves a residue of sustained noise. But we almost didn't hear it, it only registered after it was gone, and we miss it. There is a sense of loss, of emptiness.

September 8, 1999

For a long while we were stranded in the reverberations of the last imaginary composition. A new piece signals, it is exhausting to detect it. It is so adjacent to the backdrop noises of the refrigerator and the hard drive. We hear it better - it is a column of short sounds, mostly composed of clamor with roughly discernible pitches. These short sounds are very remote and placid. The noise of the drill operated by the neighbor drowns it all out. The sounds surface once again and dissolve to the left. A perception of tranquility fills the discontinuity. The sound of the car in the street does not seem to disrupt this silence. Without warning a mammoth swell of sounds made of pitches, noises, people talking, clusters of strings and rumbles of deep drums, emerges. This wave progresses, it becomes more resounding sluggishly, guaranteeing an immense smashup. But there is no eruption, without warning the wave relinquishes its motion, it just lingers in a unique spot, immobile, observing its own variance and yet it is still immutable; a massive barrier of sounds, in a single region, ever altering. Within it we hear a brief instant of brass, succeeded a long time later, by unusual mythical metallic sounding woodwinds. It remains in this solitary site for a farther drawn-out time, comparable to a burnished light, or a breeze, or a precise and tortuous carpet. Within it we concentrate on a sound of an individual in conversation with another, as if we are eavesdropping. Soon afterwards an automotive sound resembling a tractor or a tank, or a plane appears. We recognize that earlier this wall of sound started loosening. It is not fading out, or disappearing, only slackening its density. Now we hear sounds of strumming on strings, large strings, double sized strings played on the inside of an imaginary very large piano. Sluggishly an instant of the strumming surfaces then dissolves, followed by another, and another, and another. Everything stops; we don't know how we understand that this is the conclusion, the end of the piece. The noise of the refrigerator is now the noise of the refrigerator and no longer part of the piece.

October 12, 2001

7

Lonely Beating: Wittgenstein's Automaton and the Drums of War

ANAT MATAR

1

My initial question is a very familiar one: To what degree is agreement a necessary condition for understanding? You, my reader: How far do you have to agree with me in order to understand me? Any answer to this question must analyze the notion of understanding and raise such neighboring questions as: what is it that we understand? How do we understand? And how are we convinced? We all know by now that a "sentence" or a "proposition" should not be examined "in isolation"—that sentences come in contexts that are indispensable if we wish to understand them; but what does this actually amount to?

2

Traditional metaphysics relies on a notion of saturated meaning that is present to the mind's eye. The origins of meanings are conscious subjects; the meanings of their sentences represent their intentions, of which they are fully aware. A perfect understanding is then a state of total iden-

Hues of Philosophy.
Anat Biletzki (ed.).
Copyright © 2010.

tity between the origin—the writer or speaker's consciousness—and the image received in the reader or hearer's mind. Yet a serious problem immediately arises. For the subjects who are intimately familiar with the contents of their intentions, the truth of what they wish to convey forms an integral, inseparable part of the content. On that subjective level, in other words, there is no clear distinction between content and judgment at all. When we usually complain that the content of our sentences was distorted since our words were taken out of their original context, we do this because we feel that the intricacy of the saturated context was not conveyed, i.e. that the full circumstances which justify our utterances were not taken into consideration and hence the content as interpreted and received by our audience was unfaithful to our intention. A faithful citation must bring the utterance back to its origin, and the origin cannot be divided into "content" and "judgment". "If only our hearers could be exposed to our rich internal experience", we believe, "they would necessarily agree with our judgment of the content we try to convey". The upshot of this analysis is that the distinction between understanding and agreeing is blurred, if not totally eradicated, when we take seriously the traditional conception of meaning.

3

This conclusion is very inconvenient, though. Is it really the case that understanding necessarily involves agreement in judgment? A positive reply would count out, so it seems, the viability of a gradual dialogue, of a rational process which takes understanding as a necessary—and, more importantly, insufficient—first step towards agreement. If agreement is an inherent part of understanding, how can we convince each other of the truth of our judgments at all? If there is no distinction between understanding a proposition and agreeing with the claim it makes, what is left then of the notion of argument—the cornerstone of logic—and of rational discourse in general? It thus turns out that a deep tension was built into traditional philosophy from its inception. On the one hand, logical and metaphysical thinking relies on the metaphysical idea of the subject as an origin of saturated, present meaning, where judgment is an integral part of the content; on the other hand, immanent to the logical way of thinking is the possibility of argument, which sets apart content and judgment.

4

This tension subsided at the end of the nineteenth century, when Frege introduced the linguistic turn into philosophy. For Frege, metaphysical questions should be rephrased and answered with an attention to the structure of language, which makes them possible. Linguistic investigation becomes prior in the order of philosophical investigation, and this ensures—as Frege argued in the first pages of his *Foundations of Arithmetic*—that the mentalist picture of understanding is kept out of the philosophical account; any appeal to the subjective origin of meaning—to subjective experience—is ruled out. For Frege, language has an essentially logical nature. This is reflected, among other things, in the viability of the distinction between propositional content and judgment. The notion of argument is thus saved, reigning alone over philosophical discourse and shaping its nature: Agreement in judgment may come, if at all, only after understanding, grasping the content. The objectivity of content, which ensures that it is accessible to anyone, thus rescues the belief in unbiased rational judgment. This was Frege's gift to liberalism, which is, after all, the belief in such universally available judgment, which is not biased by any kind of "subjective circumstances".

5

This traditional logico-metaphysical picture has been a locus of interest for many philosophers in the twentieth century. Derrida, for one, has criticized the conception of meaning that Husserl advances in *Logical Investigations I*, where he appeals to the notion of the origin of meaning and to the idea of saturated, present content. Derrida reveals the incompleteness and rupture that necessarily penetrate into every linguistic structure which looks, *prima facie*, stable—already at the stage of the subject, the alleged "origin of meaning". Derrida no doubt follows in Frege's footsteps here. Yet unlike Frege, Derrida uses the linguistic turn in order to show that *both* traditional maxims should be relinquished: not only the saturated subjective origin but also Frege's sharp distinction between content and judgment, since it, too, relies on a stability that does not take the ruptures immanent to language seriously enough. Reading Frege with Derrida's insistence on the essential rupture of linguistic structures in mind, we realize that despite Frege's ingenuity—his success in bringing about a revolution in our philosophical methodology and his radical anti-mentalism—he too fell, eventually, into the metaphysical trap he himself had exposed. Drawing explicitly from the mathematical model, Frege proposed to analyze sentences in terms of functions and arguments, rather than subjects and concepts. That was an important step on the way to dismantling mentalism. Yet with it came the distinction between "saturated" and "unsaturated" contents—the former being independent contents, needing no completion in order to be graspable. Proper names and propositions, unlike concepts, are saturated. They refer to objects.[1] The notion of argument thus turns out to rely on an implicit appeal to the idea of saturated content, as does the mentalist picture of meaning.

6

This insight is very close to the way Wittgenstein, in his later period, conceived of the matter:

> Frege ridiculed the formalist conception of mathematics by saying that the formalists confused the unimportant thing, the sign, with the important, the meaning. Surely, one wishes to say, mathematics does not treat of dashes on a bit of paper. Frege's idea could be expressed thus: the propositions of mathematics, if they were just complexes of dashes, would be dead and utterly uninteresting, whereas they obviously have a kind of life. And the same, of course, could be said of any proposition: Without a sense, or without the thought, a proposition would be an utterly dead and trivial thing. And further it seems clear that no adding of inorganic signs can make the proposition live. And the conclusion which one draws from this is that what must be added to the dead signs in order to make a live proposition is something immaterial, with properties different from all mere signs.
>
> But if we had to name anything which is the life of the sign, we should have to say that it was its *use*.

---

[1] That is true of propositions as well as proper names, since truth-values, to which propositions refer, are taken as objects.

> ...The mistake we are liable to make could be expressed thus: We are looking for the use of a sign, but we look for it as though it were an object co-existing with the sign. (One of the reasons for this mistake is again that we are looking for a "thing corresponding to a substantive.")
>
> (Wittgenstein 1958b: 4)

The traditional metaphysical conception of language, the realist one, which takes the substantive as a model for every linguistic sign, is vital also for Frege's conception of sense. Frege's idea of the sense of the sentence is founded on the notion of saturated content, on a reference that "objectifies" it: Sense itself thus becomes, as Wittgenstein argues, something like an object. It is this conception of judgeable content which creates room for the idea of a simple argument which can be laid down clearly and unequivocally: premises, conclusion.

7

What then is the alternative suggested by Wittgenstein, that of the appeal to use as "the life of the sign"? It is, of course, manifested primarily in the style in which his *Philosophical Investigations* is written, as its author indeed declares. "The very nature of the investigation", which defies the traditional—and also Fregean—notion of argument, is connected with the book's appearance, the way it goes at persuading its reader. It is as if Wittgenstein suggests a switch from logical investigation into a kind of linguistic-phenomenology. For example, instead of analyzing the notion of understanding and directly criticizing the appeal to intentions and saturated internal meanings, he offers perceptive descriptions of the ways we feel and behave while understanding and misunderstanding. These descriptions may lead us to overcome our metaphysical urge; they may introduce, instead, a rich and intricate network of phenomena that we would like to regard as "understanding". Wittgenstein is aware that such descriptions, eschewing any appeal to the mental accompaniment of understanding, may arouse in us a strange feeling of emptiness, as if the mental is erased, is absent. He thus comments that

> ... we may feel it's plainly wrong to say that in such a case all that happens may be that I hear or say the word. For that seems to be saying that part of the time we act as mere au-

tomatons. And the answer is that in a sense we do and in a sense we don't … Just in this way we refer by the phrase "understanding a word" not necessarily to that which happens while we are saying or hearing it, but to the whole environment of the event of saying it. And this also applies to our saying that someone speaks like an automaton or like a parrot. Speaking with understanding certainly differs from speaking like an automaton, but this doesn't mean that the speaking in the first case is all the time accompanied by something which is lacking in the second case. Just as when we say that two people move in different circles this doesn't mean that they mayn't walk the street in identical surroundings.

(Wittgenstein 1958b: 157)

8

At the beginning of Israel's recent war in Lebanon, in July 2006, I found myself contemplating the question in what sense we sometimes think, understand and act like an automaton, or like a parrot. Gadi Taub, a historian at the Hebrew University in Jerusalem, and Zohar Shavit, professor of literature at the School of Cultural Studies in Tel Aviv University, declared, in articles published by the Israeli internet media, that they firmly supported the war. Ariana Melamed, a literary critic writing in the internet magazine Ynet, defended her right to hesitate, but her implicit position revealed her inclination to support the Israeli attack on Lebanon as well. The three writers had something interesting in common: They decided to express their positions through a harsh attack of those who voiced—through petitions, demonstrations and articles—an unequivocal criticism of the war from the day it began. (To be more precise, the harsh attack was directed—although this was not explicitly mentioned—only towards those *Jews* who opposed the war; this point of precision will turn out to be of great importance later.) All three writers chose, moreover, the term "automatic" in their characterization of those who resisted the war.[2] My immediate reaction upon reading those assaults was a double

---

[2] Taub (*Ynet*, August 13, 2006, in Hebrew) writes at length, in consequent articles, about "this degenerated and automatic kind of leftism." Shavit (*Ynet*, July 31, 2006, in Hebrew) claims that there is, in Israel, "a group of people from the radical left, whose support of

lack of surprise: I was not surprised by the fact that the threesome regarded Israel's decision to wage a war against Lebanon at that time as a balanced and morally-justified one; and I was not surprised, also, by the accusation of automatism that was leveled against me and my fellow-objectors. This double lack of surprise invites philosophical reflection. It is, I wish to argue, closely connected to the question of the temporal and conceptual relations between "understanding" and "judgment" and to Wittgenstein's insights regarding the experience of meaning.

9

We may try to give a finely tuned phenomenological description, Wittgenstein-style, of some Israeli citizens' reactions to their country's decision to wage a war against Lebanon, following the events that occurred on the Lebanese border on July 12, 2006—when eight Israeli soldiers on patrol were killed and two others were abducted by Hezbollah warriors. On one side there were those who believed that Prime Minister Olmert, Minister of Defence Peretz, and Chief of Staff Halutz should be "given a chance"; those who felt the war was a just and moral one, a "proportional reaction" to Hezbollah; those who blamed the left of "automatism", and eventually, after about three weeks, declared that it was time to stop. It was remarkable to notice, for example, their unquestioning acceptance of and uncanny joy in repeating the horrific descriptions of the bleak future awaiting us all if Hezbollah and Iran are not stopped immediately. There was also the joint amnesia—a sweeping forgetfulness of what actually happened during the first day of the war, e.g. of not even one missile having been launched on Haifa or Karmiel before Israel attacked Beirut. And most of all, perhaps, there was the craving for patriotism, for a whole-

---

the other side is automatic." From her point of view, the Israeli army's action in Lebanon "is not cruel and is not clumsy and is not exaggerated and is not extreme." Ariana Melamed (*Ynet*, July 24, 2006, in Hebrew) insists on her right to "a perspective of a perplexed leftist"; she wants to just "keep cleaning the bomb shelter near the house." The fact that she withholds judgment is presented in her article as a virtue, a much better reaction to the circumstances than that of those who "are not at all confused by flying missiles and roaring cannons," those who do not dare examine basic presuppositions and adhere to a "sharp, focused, black-and-white" worldview. (These adjectives call for future attention as well.) The unperplexed stand side by side "in an obedient line of sheep who rehearse as a chorus what was so obvious that they would say."

hearted, unqualified support of the homeland.[3] We may compare these reactions to the other side, to the experiences of those who opposed the war from its very first moment, those who were overwhelmed by feelings of despair and nausea, as the news came pouring in about incessant bombshells over the whole of Lebanon, about dozens of Lebanese citizens killed and hundreds who became refugees. Here there was the bitterness and rage of impotent men and women who knew—automatically, without doubt—that soldiers would die in vain, that the army would not function, that no one would take care of the home front.

The way we experience meanings is gained very slowly, by training: It is shaped by our parents, at the nursery, in school, and through formative events in our life. The differences resulting from different trainings are significant. Think for example of different possible reactions to seeing a soldier, or better, an armed border-policeman, in uniform, walking in the streets of Tel Aviv. One may feel abhorrence and resistance, fear or respect, or perhaps nothing special, taking the soldier's presence and appearance as a matter of course. Think of the various possible experiences when some relative—a brother, a nephew, a neighbor—enlists in the army: some people happily escort the boy to the induction center, take pictures to commemorate this important moment, and are full of good advice (and nostalgia), reflecting on their own army service. Some, on the other hand, feel sorrow and estrangement, hoping that the inductee will soon realize his mistake and decide to quit ...—As Wittgenstein said, two people may move in different circles and yet walk the street in identical surroundings. These "experiences of meaning" are, of course, devoid of any value if they are not supported by strong factual and argumentative accounts, by historical data and by political analyses. However, aren't these rational accounts and analyses meaningless without the emotive foundations depicted here? Our judgment—for instance, of the official justifications of Israel's actions in July 2006—depends upon the way we experience the meanings of these justifications, and this way is not easy to convey in an argument: premises, conclusion.

---

[3] Prominent Israeli author A. B. Yehoshua expressed this craving explicitly. "At last", he wrote, "we've got a just war" (*Ha'aretz*, July 24, 2006).

10

And this, indeed, brings us back to Wittgenstein's *Brown Book* and to his comments on understanding a sentence, and on the illusion of the metaphysical—but also Fregean—conception regarding such understanding:

> What we call "understanding a sentence" has, in many cases, a much greater similarity to understanding a musical theme than we might be inclined to think. But I don't mean that understanding a musical theme is more like the picture which one tends to make oneself of understanding a sentence; but rather that this picture is wrong, and that understanding a sentence is much more like what really happens when we understand a tune than at first sight appears. For understanding a sentence, we say, points to a reality outside the sentence. Whereas one might say "Understanding a sentence means getting hold of its content; and the content of the sentence is *in* the sentence".
>
> (Wittgenstein 1958b: 167)

The Fregean conception of language does not *really* give priority to the sentence over what it refers to, outside of language. Immanent to it is the thought that words essentially conceal, that they refer to something which hides behind them, to the "thing corresponding to a substantive". It is that external *thing* that is important, rather than the words. Wittgenstein's claim is that this conception stands in contradiction to the spirit of the linguistic turn. In the case of music, on the contrary, we do not tell the expression apart from the feeling that it arouses in us.[4]

I now wish to appeal to this Wittgensteinian insight in order to make clear what it is that I am actually driving at. When Israeli national songwriter, Naomi Shemer, died, in June 2004, Ariana Melamed wrote an animated obituary:

> And when we hum the tune of "We Are Both from the Same Village", when we are swept by her "Chariots of Fire", or when we wrap ourselves with sweet sadness listening to

---

[4] Wittgenstein elaborates this idea in his *Philosophical Investigations*, part II, paragraph vi.

> "Hot Days in the Military Post", when we pray like devout agnostics and ask our Good Lord that the "rooted not be uprooted",[5] we repeat and recite, full of magic, the key sentences of Israeliness, that evasive thing that has no definition but which turns us all into members of one culture, rather than scattered tribes. Naomi Shemer's words have this uniting power in them.

This precisely is my point. The idea that Melamed fully understands here—the importance of that which evades definition but shapes the discourse and actually makes it possible, *the latent conviction that precedes understanding*—is forgotten in her article about the war. The uniting power of Shemer's nursery rhymes (and this perhaps is the moment to note that all her songs are eventually nursery rhymes) is the power to form an (automatic?) alliance. One of the most salient characteristics of her lyrics is that their focus is limited to the Israeli-Jewish experience, yet their message is always a purportedly universal one: a saccharine perspective of an Israeli-Ashkenazi-middle-class-Jew that is generalized in total blindness to the fact that this experience is rather particular. "We" are "all" members of "one culture"—the "Israeli culture" (that has nothing to do with any Arab, naturally; and as a matter of fact, with any less-bourgeois Israeli Jew as well.)

11

The naïve liberal belief in the viability of non-automatic thought, which may be fully articulated and voiced, is intimately connected to such a self-authoritative-universal perspective. People adopting such a perspective move in the circles which shape their form of life: This is the form of life of the privileged, the safe and satisfied. An essential ingredient of this perspective—an element that is virtually constitutive of it—is its blindness to the fact that theirs is simply a form of life, i.e. that it is one

---

[5] All these are titles and lines from Shemer's lyrics. Melamed published her obituary in *Ynet* on April 26, 2006. English readers may get a better grasp of my claim by comparing Melamed's "we" with the attitude expressed by Eric Silver in the obituary he published in *The Independent* on April 30: "Naomi Shemer wrote the kind of songs Israelis wanted to hear: warm, patriotic, nostalgic, rooted in childhood memories and an idealized biblical landscape... Soldiers were comrades, never oppressors. The land 'belonged' to the Jews."

among many others. That same blindness yields their conceptions of the notion of understanding and of the notion of argument that goes with it—and also, of course, their conviction that these conceptions are not biased in any way. Such blindness, I wish to emphasize again, is a necessary ingredient of the liberal form of life.

12

Am I arguing, then, that logic, argument, or indeed facts, history, analysis, understanding, are only myths that should vanish in thin air? Not at all. The *argument* I advanced here—and I take all twelve stages as necessary: There was no deviation—leads to the conclusion that every argument depends on practice, on lived experience, which cannot be reduced to non-emotive discourse. Shavit, Melamed and Taub disregarded the fact that their judgment is as automatic as mine, and that it is strongly connected to their particular perspective, which we may dub "the Naomi Shemer perspective" rather than, for example, to the perspective gained by a persistent practice of political activism and active exposure to the form of life of the under-privileged. A substantial exposure to the form of life of the Arab citizens of Israel, for instance, would rescue the warmongers from the shame of their July 2006 articles. (How many Israeli Palestinians were "perplexed" like Melamed at the beginning of that dreadful war? How many couldn't decide whether it was a justified war or not? And what can be learnt from this? Is this population much more automatic than Professor Shavit? What is the connection between the answers to these questions and the answer to the question how many Israeli Palestinians share Melamed's elation when they hear Naomi Shemer's tunes? And is there any connection between the answers to all these questions and the fact that, in their articles from the first days of the war, Shavit, Melamed and Taub totally forgot about the existence of a fifth of the population of Israel?) Those who do not share, albeit partially, the experiences of people who are constantly exposed to the violence of the army and police forces; those who never witnessed the border-police's brutality in the more wretched neighborhoods of Tel Aviv or near the Apartheid Wall—how on earth could they develop a healthy revulsion towards their uniforms? How could they perceive reality in a way radically different from the one they were trained for in the nurseries and schools of the privileged? How could they overcome their natural tendency to hum and wrap themselves up with the self-adoring sadness of the "Israeliness" which shapes their views?

13

Resisting the temptation to reduce the concrete and political to timeless formulas must not end up in logo-phobia. Indeed, intuitions and experiences of meaning are never enough for judgment, as we have noted earlier. Detailed accounts—historical, personal, aesthetic, political, economical—and analyses, theories, hypotheses and beliefs are no less necessary. (Do I really have to mention that the radical left activists who "automatically" resisted the war from its first day explained their opposition by appealing to statistics, analyses, alternative descriptions of the situation and comparisons with past experiences? And on the other hand, given what we now know about the political options that Israel had in June 2006 (e.g. from the Syrian mediator Ibrahim Suleiman)—did any of those who supported the war when it started change his or her mind?)

Thus, this article does not call for doing away with argument, or indeed, philosophical discourse. What it does do is urge us to remember the fact that thought is colored by experience, and that this experience is often political. This is true in particular of traditional philosophical thought about thought itself, about argumentative philosophical discourse. Philosophers tend to forget how much their own ways of formulating and arguing are invested with what seems to them obvious, and how much this "obvious" is context-dependant—the context being not only historical, but also sociological and personal. Nietzsche's criticism of Kant, for example, may be read along these lines.

14

Although he did, of course, criticize philosophers for ignoring their own dogmatic blind spots, Wittgenstein never emphasized the political aspect I discuss here. Yet the following story may tell us something about the political implications of philosophical thinking. In a letter he wrote to Norman Malcolm, Wittgenstein reacted to a "primitive" comment "of national character" that Malcolm made. The comment shocked Wittgenstein:

> I then thought: what is the use of studying philosophy if all that it does for you is to enable you to talk with some plausibility about some abstruse questions of logic, etc., and if it does not improve your thinking about the important questions of everyday life, if it does not make you more conscientious than any journalist in the use of the *dangerous* phrases such people use for their own ends? You see, I know that it's difficult to think *well* about "certainty", "probability", "perception", etc. But it is, if possible, still more difficult to think, or *try* to think, really honestly about your life and other people's lives. And the trouble is that thinking about these things is not *thrilling*, but often downright nasty. And when it's nasty then it's most important.
>
> (Malcolm 1958: 39)

It seems then that Wittgenstein still clings to a remnant of the conception I wished to criticize by using his notion of the experience of meaning and his criticism of Frege's failure to implement the linguistic turn. Wittgenstein clearly separates "abstruse questions of logic" from "the important questions of everyday life"; yet he does make it just as clear that for him philosophy has little value if it does not make one more politically conscientious.

*****

This volume is dedicated to the memory of my beloved friend, my first philosophy teacher, Ruth Manor. Like Wittgenstein, Manor separated the logical from the political; and like Wittgenstein, she had a sensitive ear for manipulative, dangerous phrases. Most importantly, she hated, quite automatically, the drums of war.

References

Malcolm, Norman (1958). *Wittgenstein: A Memoir*. Oxford: Oxford University Press.
Wittgenstein, L. (1958a). *Philosophical Investigations*, trans. G. E. M. Anscombe. Oxford: Basil Blackwell.
Wittgenstein, L. (1958b). *The Blue and Brown Books*, 2nd ed. 1969, Oxford: Basil Blackwell.

8
Living with Paradoxes
ANAT BILETZKI

1 Introduction

Let me begin with Manor's own words. In 1983 she published a thought provoking paper "What is paradoxical about paradoxes?", saying:

> ... [T]here is almost no discussion of the general phenomenon of paradox, or suggestions for a general theory which will characterize and explain this concept.
>
> This is quite a surprising fact. The concept of paradox is a logical concept related to certain kinds of arguers and any rational theory dealing with "correct" or "incorrect" arguments must elucidate this concept as well.
>
> (1983: 249)

She went on to enumerate a number of questions that a theory of paradox would need to answer:

> 1) What is paradox, what is a solution to a paradox, and what is the difference between solving and evading a paradox?
>
> 2) What is the relationship between paradoxes and theories of rationality and logic? In particular, how can the phenomenon of paradox be explained in the context of theories of rationality that include a strict requirement of consistency?
>
> 3) What is the importance evidenced in paradoxes and what are the methodological uses of the concept of paradox?
>
> (1983: 249-50)

There were three propositions that Manor was, ultimately, intent on defending:

> 1) The supposition that every paradox is, in principle, soluble is problematic to the same degree as the contrary supposition.
>
> 2) There are good reasons to suppose that there are paradoxes that are, in principle, insoluble.
>
> 3) Whether there are such in-principle insoluble paradoxes or not, we often find ourselves in a situation where we subscribe to a theory, in which a paradox crops up that we have (maybe only meanwhile) no idea how to solve. Thus, we must explain how this might happen and how we are to behave rationally in such a case.
>
> (1983: 250)

Manor went on to submit a logical-analytic-methodological discussion, where she elaborated on the difference between extensional and intensional definitions of paradox, on their syntactic and semantic traits, on alternative logical systems that are (good?) candidates for paradox-solving, and on the distinction between "solving" a paradox and "evading" it.

> In general, a solution to a paradox shows that the decision [to choose between accepting the "good" argument in a paradox, or rejecting it] is unnecessary since there is an error in the considerations leading to it. In general, evading a paradox means limiting language so that the paradox is no longer given to formulation and it is therefore impossible to reach a point where a decision, this way or that, must be made.
>
> ... [A] solution to a given paradox must be a theory that fulfills the following conditions:
>
> 1) The theory must explain the fallacious steps in the paradoxical argument.
>
> 2) The theory must explain why we made those fallacious steps. In particular, if it transpires that we used a certain rule of inference incautiously, the theory must characterize the conditions under which use of certain rules is valid and the conditions under which it is fallacious.
>
> 3) The theory must allow us to understand the contradiction without committing it ourselves. In particular, we must be cautious of holding on to one branch of the argument as a claim against the correctness of its other branch ....
>
> 4) The theory must have independent rational justification, independent of the considerations in the paradox itself, and not ad-hoc.
>
> (1983: 258-59)

And she culminated the presentation, for the while, in her usual, distinctive style:

> ... [A] paradox presents us with a dilemma: either we accept the proven contradiction and "learn to live with it", ... or we give up on the customary inferential procedures that led to contradiction. A solution of the paradox attempts to pass be-

tween the two horns: not to accept the contradiction and not
to completely give up on customary inferential procedures,
but rather to point to the fact that, in our considerations, we
took steps that are unacceptable and thereby fell into fallacy.
The theory that we need as a solution is a comprehensive
theory which will justify the valid deductive procedures that
we used and explain the fallacious ones that we used in the
paradoxical argument. If we don't have such a solution, then
it is not clear, a priori, why we shouldn't learn to live with
the contradiction or why we should not give up on our customary inferential practices (and turn to sport or politics).

(1983: 259)

I will here attempt to investigate Manor's almost by-the-way option—that
of living with paradoxes. I will claim that there are, indeed, insoluble-in-
principle paradoxes and that these are the logical paradoxes *par excel-
lence*; they are, indeed, more than contradictions. If this is the case, and if
there is an intimate connection between logic and rationality, then, given
that real paradoxes are a problem for and in logic, they are then a problem
for rationality. It then appears that it is somewhat more difficult to live
(rationally) with paradoxes. Or perhaps one must expand the concept of
rationality—so that it will treat of more than just "correct" and "incor-
rect" argumentation—and, in turn, of a rational life, including common
sense and even morality.

2 What Is a Contradiction? What Is a Paradox?

Aristotle taught us of contradictions: two propositions contradict each
other when they cannot be true together and cannot be false together. A
little less formally, we conventionally put it that A and not-A, a proposition and its negation, contradict each other. That same Western, logical
tradition, oft-times explicitly termed logocentric, also learned something
else from Aristotle—the law or principle of non-contradiction: contradictions cannot exist, or, differently put yet again, both sides of a contradiction cannot co-exist. But what is the import of that "cannot co-exist"? At
this point, and following Aristotle's three versions,[1] the interpretations of

---

[1] Famously, Aristotle formulated three versions of the law of contradiction—1) the metaphysical (ontological) law: "It is impossible for the same thing to belong and not to be-

the law of contradiction start diversifying. Some say that a situation and its denial cannot obtain simultaneously "in the world"; i.e. a metaphysical, even ontological reading of the law of non-contradiction. Some consider this is a law of thought—we cannot think, simultaneously, at the same time-place, in the same sense and manner, the content of a proposition and its negation. A third option, a more popular one in these post-linguistic turn days, when meaning has become the fulcrum of discussion, focuses on language: one cannot simultaneously state a proposition and its negation. Although these are essentially different readings of the principle of non-contradiction, whichever is the correct or chosen interpretation, the metaphysical, cognitive, or linguistic version, we ultimately reject the contradiction. And this is of note: the contradiction is perceived negatively, as something to be shunned, as a situation, a thought, or a proposition that is intolerable. It is the source and origin of something very "wrong"—perhaps not of all that is wrong, but definitely one source of identifiable wrong—in human thought. Wherefrom such negativity? Why do we not even entertain the feasibility of contradiction?

The logical–scientific reply to this question is facile: Using that same logic which is constituted by the law of non-contradiction we can show that anything can be inferred from a contradiction. Since we are in pursuit of "true" theories, the viable option of proving everything from a contradiction empties the concept of "truth" of all content. Given a contradiction, anything can be proved. For the scientifically inclined, for the seeker of (exclusive) truth, this is debilitating. Just as challenging, however, is the "atmosphere" surrounding contradictions, an atmosphere that emanates from our incorrigible demand for a rational life. What we encounter in the Western tradition is revulsion from contradiction and an attachment to consistency—the absence of contradiction—in order to preserve human rationality. (The logical-scientific aspect of our anti-contradiction attitude and the rational one are not unrelated; quite the contrary.)

What of paradox? What is exclusive about paradox? What makes it more interesting, more fascinating, more elusive, more fruitful, more exercising than "just" contradiction? What would be a definition of para-

---

long at the same time to the same thing and in the same respect" (Aristotle, *Metaphysics* IV 3); 2) the doxastic version: "It is impossible to hold the same thing to be and not to be" (Aristotle, *Metaphysics* IV 3); and 3) the linguistic (semantic) principle: "Opposite assertions cannot be true at the same time" (Aristotle, *Metaphysics* IV 6).

dox which could differentiate it from plain contradictions? It is obvious, of course, that paradox has something of the contradiction in it; but it also has an additional something that endows it with the label of paradox, one-over contradiction. What is it?

The professional literature—and speaking of professional I allude here to formal logic, in which most specialized attempts at definition have been made—suggests various characterizations, sometimes even definitions, of paradox. Without perusing these in (mostly formal) detail, we note that they all turn, first and unsurprisingly, to contradiction as inhering in paradox. But these are special contradictions so, in pursuit of the paradoxical side of these unusual contradictions, some turn to the logical *necessity* of the contradiction. (This is the strategy employed in the, apparently, most natural definition, which sees paradox as an argument with true premises and a valid inference leading to a contradictory conclusion—the definition usually employed by Manor); some emphasize the *self-referential* aspect of paradox (mostly accruing to the most renowned paradoxes—the liar, the set, the barber); some point us to the *surprising* element in the contradiction (subsequent to our complacency about the logical steps that led to it); and some address, more forcefully now than in "regular" contradictions, the aspect of *negation* in paradox (since we cannot really produce a paradox without basing its formulation on negation—as in, again, the liar, the set paradox, etc.). Putting these all together, accepting them as the intuitive conglomeration accompanying the concept of paradox, one arrives at the distinction between mere contradiction and paradox. When faced with a contradiction and recognizing it as such, we are committed to cancel out one of its horns; we are compelled to decide on either A or not-A. What is interesting, fascinating, and unrelenting about paradox is precisely the impossibility of opting for either of its sides by cancelling out the other. This is because there is—between the contradictory ends of a paradox—a logical, necessary, dynamic (some will even insist on "dialectical") connection that prevents us from choosing one over the other. The problem, and the magic, in paradox is exactly that, having chosen one side of the contradiction, we are thrust, immediately, unwillfully, choicelessly, to the other. This is not a matter of A and not-A statically facing each other , both being scrutinized from an external vantage point, so to speak, by someone—a philosopher, a scientist, even a religious person—who can then choose between them. This is rather a compulsory flow between A and not-A that makes it impossible for one to stand aloof. Accepting A, she is led, perhaps unintentionally, to not-A; and then back.

3 A Role for Paradox (and Contradiction)

The above articulation of paradoxes and contradictions is logically-oriented. These definitions, this roster of characteristics, and this kind of treatment of paradoxes and contradictions are conventionally placed and are, in fact, popular in the discipline of philosophy in general and, even more so, in the field of formal logic in particular. It is in logic that we purport, first, to be able to identify and recognize contradictions and paradoxes. Secondly, philosophy pretends to be able, sometimes with the aid of logic, to solve the paradoxes. We do not here speak of solving the contradictions; a contradiction is not a problem to be solved. A contradiction, we have seen, is an ontological, epistemic or linguistic phenomenon which must be identified as such. (And a debate is nothing but the human context, the praxis wherein human beings position themselves on either side of a contradiction and conduct a discussion, a debate, in order to choose one side or the other. In this sense, and as strange as it seems, one may say that contradictions are simpler than paradoxes.) We either recognize contradictions with despair as here to stay or take a step, in debate, towards this or that side of the contradiction.

In "real" paradoxes the scene—and it is not a scene of debate—is different: philosophers, and logicians, face a contradiction that will not desist, a contradiction that will not allow a choice between its sides, since choosing one leads to the other or both are logically mandated. In other words—and here I am propounding a personal opinion, albeit a personal logical opinion—a real paradox is neither dismissable nor soluble. And the question, which is the more interesting question, then becomes— what is to be done? Three suggestions come to mind.

3.1 The Religious-Mystical Option

In the context of religion one significant yet relatively uncomplicated point, having to do with contradiction, must be recognized. It is sometimes claimed that the shackles that have been put (or just happen to be there) on human thought, the restrictions that obligate us when we aim at thinking coherently, consistently, or even rationally—including, certainly, the law of non-contradiction and the absolute denial of contradiction—are not compelling in the religious context, i.e. in godly deeds, in godly thought, and in the speech of and about God. In other words, the requirement to abstain from contradictions, which regulates human thought and speech, is nonexistent in our talk about God, about his words and his deeds. This insight can be straightforwardly couched in the following elaboration: Instead of speaking about the exclusivity of God's words and deeds as lying precisely in their independence of the law of contradiction, the discussion can, and perhaps should, turn to *our* understanding of God, *our* reportage of his talk, and *our* interpretation of his deeds. Then, since human understanding is not free of the law of contradiction, that is to say, if we cannot understand that which is not bound by the law of non-contradiction, it follows that if God (and his words and deeds) are not ruled by the law of non-contradiction we cannot fathom them. Put differently again, the turn to contradiction in order to *explain* that which is *inexplicable* about God is fraught with pit-holes, paradoxical holes. These can be embraced, not evaded, by admitting that we cannot explain, and therefore cannot understand God precisely since the godly is not ruled by the law of non-contradiction. God is a paradox.

3.2 The Wittgensteinian Option

Space and time do not suffice to expand here on Ludwig Wittgenstein—on his convoluted (according to some), spectacular (according to others) thoughts of mathematical contradictions and paradoxes, on his enigmatic thoughts on religion, and on his unusual philosophy and its place in the Western philosophical tradition. Some scholars have located the religious option above—which speaks to the inability of contradiction to reside in human thought and language—in Wittgenstein's claim that religion inhabits the mystical place which is beyond the limits of language and world.[2] I look to pinpoint a different Wittgensteinian place. But first—two notes:

In his philosophical work, Wittgenstein emphasizes and prefers ordinary language, as opposed to high-brow philosophical language;[3] similarly, he favors the deed, as opposed to the word, the thought, the logos. Quoting Goethe, he agrees that "in the beginning was the deed" (Witt-

---

[2]"God does not reveal himself *in* the world" (Wittgenstein 1922: 6.432).

[3]This is made explicit in the later Wittgenstein and is arguable, dependant on interpretative positions, concerning the early Wittgenstein.

genstein 1969: §402). Secondly, Wittgenstein tells us that our fear of contradiction and paradox is bogus. We find ourselves in a cultural framework that has, for reasons of its own, inculcated us with lack of tolerance for contradictions. Speaking thus, Wittgenstein succeeds in arousing against him both logicians and, mainly, mathematicians who insist on the rejection of contradiction as a fundamental element in our basic conceptual scheme.

What must we do, then, what *can* we do, according to Wittgenstein, with contradictions and paradoxes? He submits that "[w]e might ask: what role can a sentence like 'I always lie' have in human life?" And he replies modestly but confidently: "... here we can imagine a variety of things" (Wittgenstein 1967: V-30). Indeed, there are various circumstances about which Wittgenstein challenges us. "For might we not possibly have wanted to produce a contradiction?" (Wittgenstein 1967: II-81). Not surprisingly, in the Wittgensteinian context of meaning-as-use, if we can identify a use for contradictions and paradoxes, they lose their aura of meaninglessness. "Perhaps we should say of this man [who says I always lie] that he means perhaps something like: What he says flickers, or nothing really comes from his heart" (Wittgenstein 1967: III-58). Or "... change might be expressed by means of a contradiction" (Wittgenstein 1967: V-8). In other words, we can live with contradictions and paradoxes if we understand the role they play in our life.

3.2. The Religious-Wittgensteinian Option

"I am not a religious man," says Wittgenstein, "but I cannot help seeing every problem from a religious point of view" (Rhees 1984: 79). And he goes on to describe the religious person as playing an utterly different language-game from the secular person's game. This is a language-game in which contradictions and paradoxes can be accepted without disgust; and so strong is this option that it permits of a different definition of rationality itself. This language game is not dependent on and does not belong to the scientific language-game, the mathematical language-game, the logical language-game, or any other language-game in which we are prone to find definitions of rationality. As is usually recognized, these comments on the religious language-game are intimately related to the concept of "form of life", also used by Wittgenstein to refer to a religious form of life in which the meaning of words we use and the demands we make on concepts are different from those in other forms of life—most evidently, the secular form of life.

Now, ordinary, ever-day language is the framework that allows for our contact with paradoxes. How? In praxis, and most particularly in human religious praxis. There is no necessary partition, as we might have demanded in very systematic theology or very outlandish mysticism, between a godly language which tolerates contradictions and human language which does not grasp them. The existential paradox consists precisely of human language, which is the human form of life, which can and cannot understand contradictions, and which thereby countenances paradoxes.

How can this be understood analytically? No need, says Wittgenstein, to engage here with analysis. We should only gaze—"look, don't think" (Wittgenstein 1953: §66)—at traditional-religious writings and deeds in order to penetrate this form of life, and thereby to live in and with its paradoxes.

4 Back to Manor

> Dilemmas, contradictions, and paradoxes are a daily occurrence in our life. The attitude, represented by classical logic, of safeguarding consistency at any price stops where inconsistency is discovered and does not explain how to go on rationally.
>
> (Manor 1983: 262)

> Whether there are insoluble paradoxes or not, we must learn to live with them since we do actually live with them, in the sense of encountering paradoxes for which we have (as yet?) no solution …. For that we need a theory of rationality that will explain how paradoxes are created and how we can live with them. Concerning the first point, our theories, formal or otherwise, are supposed to describe a full spectrum of primary intuitions so it is no wonder that they often conflict with one another …. [Concerning the second point,] the logician must provide … a system which we can continue to use even if we encounter a paradox or a contradiction.
>
> (1983: 263-4)

> Finally, the "interesting" paradoxes, those that have no solutions or that have too many solutions, represent the no-man's land between the different theories. On the one hand, they point to the fact that the demand for consistency as a basic requirement in theories of rationality has qualifications; and, on the other hand, they show that in several fields there are open questions which have not yet been solved. The chance of finishing all the work and suffering boredom is slim.
>
> (1983: 259)

This is Manor's essential optimism. It is an optimism which holds, within itself, a type of contradiction: the desire and the ability to continue doing logical-philosophical work in solving paradoxes (and thereby never being bored) with the understanding that we "live with paradoxes". However, the optimism itself is born of the conviction that the paradoxes with which she deals have no solution "*as yet*", but that there is reason to carry on in pursuit of such solutions. In another article, "Simulating Imagination" (1984), Manor goes on to claim that it is precisely our encounter with paradoxes that is responsible for "creativity" since it is in the context of that encounter that we attempt to find original solutions to impossible situations. Indeed, several psychological theories attest to the constructive character of contradictions in our life; aesthetic theories evidence the fruitfulness of perceptive contradictions in art; and even political theories show our behavioral contradictions as leading to effective and constructive conflicts and tensions. This is a family of views and attitudes that accepts the "impossibility" of contradictions and paradoxes as a needed presupposition and then enjoins us to continue the search for solutions. The chance that we'll finish the work—the logical, philosophical work—and suffer boredom, is slim. In other words, we must live with paradoxes by continuing to look for solutions. Truly, an essential optimism.

To my mind, however, this is a view contrary to really, authentically "living with paradoxes"; I opt for another claim. "Real" paradoxes are, as noted above, a necessary consequence of our logic. These are insoluble paradoxes, not as yet but forever, precisely because we are imprisoned in our logic, in a certain mode of thought. The question then becomes—how can we live with them logically? How can we countenance such a logic? If we cannot take the analytic path to solving paradoxes and do not choose the mystical way (which tragically accepts contradictions or unequivocally rejects all common sense), if we do not think it suffices to simply describe—in Wittgensteinian manner—the paradoxical form of life, how can we truly live with paradoxes?

A preliminary intimation at a reply will suffice here: rationality is not exclusively, or even necessarily, logic. Rationality, in the humanistic sense of reasonableness, includes far more than logic; it constitutes an essential tool with which to manage our life. If real paradoxes are a direct, absolute, and necessary consequence of logic, if we cannot dismiss or solve them, and if acknowledging this comprises a constructive stage in our life, then a rational life cannot, subsequently, be simply logic-dependent. Not only is the requirement of consistency as a basic demand of rationality thereby qualified and limited, but a basic demand of a rationally lived life is to live with paradoxes, consciously and with acceptance. This is, perhaps, the opposite of Manor's optimism. However, even if we give up on (classical or formal) logic as the exclusive attribute of rationality, we do not relinquish reason, common sense, and morality as fundamental conditions of a rational life. In other words, perhaps despair—and the particular despair of solving logical paradoxes—is not, as is usually thought, an irrational emotion but is rather the reasonable consequence of our paradoxical life.

References

Manor, Ruth (1983). "What Is Paradoxical about Paradoxes?" In *Israeli Philosophy*, ed. by Asa Kasher and Moshe Halamish. Tel Aviv: Papyrus, pp. 249–72.

Manor, Ruth (1986). "Simulating Imagination". *Logos* 7, pp. 69-81.

Rhees, Rush (1984). *Recollections of Wittgenstein*. Oxford: Oxford University Press.

Wittgenstein, Ludwig (1922). *Tractatus Logico-Philosophicus*. Tr. C.K. Ogden and F.P. Ramsey. London: Routledge.

Wittgenstein, Ludwig (1953). *Philosophical Investigations*. Ed. G.E.M. Anscombe and R. Rhees, tr. G.E.M. Anscombe. Oxford: Blackwell.

Wittgenstein, Ludwig (1967). *Remarks on the Foundations of Mathematics*. Ed. G.H. von Wright, R. Rhees and G.E.M. Anscombe, tr. G.E.M. Anscombe. Oxford: Blackwell.

Wittgenstein, Ludwig (1969). *On Certainty*. Ed. G.E.M. Anscombe and G.H. von Wright, tr. D. Paul and G.E.M. Anscombe. Oxford: Blackwell.

www.ingramcontent.com/pod-product-compliance
Lightning Source LLC
Chambersburg PA
CBHW071509150426
43191CB00009B/1460